隨心。煮意

Cook it My Way

Gigi Wong _ 黃淑儀

美食無邊界 # 煮自己喜歡的味道

序 _ 鄧達智

孜孜不倦黃淑儀

不貪圖享受日出日落家庭主婦悠閒。子女成長爾後，無線當年四大花旦、全港首套長篇電視劇「夢斷情天」女主角黃淑儀重出江湖，近年拍劇不停，主持美食節目不拙，還有廚藝書籍一本接一本不斷煮、不斷著作，點止演藝人咁簡單？

Gigi 姐着實才女一名！

廚藝書籍簡潔大方，易學不流粗糙；文字簡化，程序易達。原籍香港、留加留美腦幹細胞科學家、北京大學教授于常海是 Gigi 姐頭號粉絲，揹着她筆下好大一套廚藝書籍在北大校園如魚得水，按書指示，一步一步將黃教頭的妙手生花展示飯桌；少年學生、校園同事、教授長老都被于教授廚藝降服，吃過常問：于教授甚麼時候又開餐？

不少人印象：「廚書一般好睇唔好使，跟足程序，煮出一鑊粥（搞砸了）！」

跟隨黃淑儀，結果教授成為廚藝萬人迷。

有幸被邀跟 Gigi 姐在其節目上出演共煮嘉賓，從旁得觀廚藝達人不簡單，無分大小每個細節一眼關七，卻又從容不迫；每一步驟看來不需咬牙切齒用大力，而是輕於鴻毛似地恰到好處。

每次出鏡前，Gigi 姐都留給自己足夠時間，細讀將要拍攝的內容，也再讀原本是她自己寫的食譜，關鍵在於要配合烹調的程序、配合鏡頭的運用和配合導演的要求，縱使煮她自己著作內的菜式，她也不會掉以輕心；猶似私底下我們十分熟悉的 Gigi 姐，工作、行為、言語嚴謹而態度平易近人，更不恥下問，經常請教各路高手，鑽研更高、更深、更廣博的廚藝。

自序 _ 黃淑儀

第 13 本書的感言

近年來，出外拍攝烹飪節目的機會多了。

除了中國大陸各省各縣……南至海南島，北至哈爾濱，台灣的嘉義、台南與台北，東南亞的新加坡及馬來西亞，甚至南半球澳洲的悉尼、墨爾本及亞得萊德……

增廣見識，眼界大開！

除了認識更多新穎食材，學識更多新鮮配搭，更印證了各師各法的烹調方式。

不能説哪一個地方的食物更美味，只能説哪一個地方的烹調方法更符合你的要求，哪一個地方的口味更接近你的味蕾或童年回憶！

我不勉強自己拋棄舊有的點子，只能在食材上加多一些些，烹調時改變一點點。

畢竟，袋鼠尾不易買到，我們也只有用回牛尾！

又或者，內地的粉葛肉嫩多汁，可以燜五花腩，而我們買到的，是乾身又多渣的粉葛，那就只能用作煲湯了！

不要嘗試改變自己的風格，只求在突破中有進步的空間！

目 錄 _CONTENTS ▽

cold dish_snack　rice_noodles　main dish　soup　sweet

cold dish_snack

rice_noodles

小菜 main dish

湯 soup

甜食 sweet

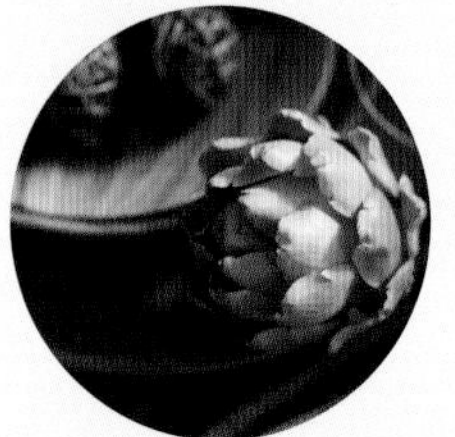
cold dish_snack

rice_noodles

main dish

soup

sweet

cold dish_snack

rice_noodles

main dish

湯 soup

甜食 sweet

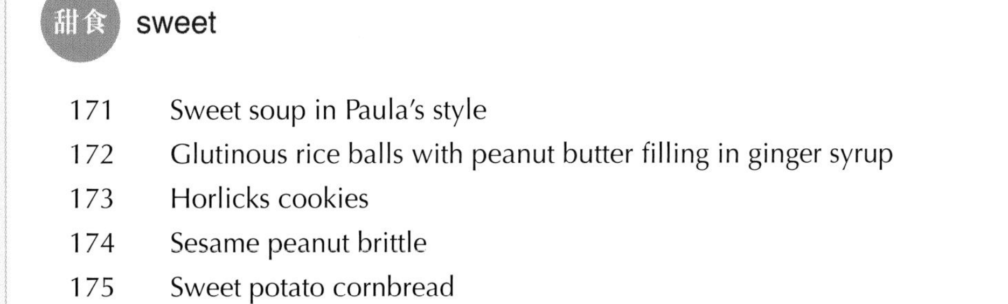

冷盤小吃 cold dish_snack

涼拌麻辣雲耳

Gigi Wong 麻辣是亮點 ♥ 開胃菜 ✽ 味蕾的新挑戰 ♨

可將醬汁和雲耳一起倒入密實袋內，鎖實，搖晃十數下，待醬汁和雲耳完全拌勻即可倒出享用。

雲耳

食譜採用俗稱老鼠耳的雲耳。它在浸發後比一般雲耳小，處理時毋須再撕成細朵，省卻工夫。

經冷藏的麻辣雲耳，入口冰涼，口感爽脆，舌尖有麻香，是刺激的夏日消暑菜。

材料 @ 雲耳……1 杯 #浸軟，剪去蒂，汆水 5 分鐘，待涼

醬汁 @ 麻辣鮮露……1 湯匙
蠔油……1 湯匙
糖……1 茶匙
蒜茸……2 粒 #剁碎
芫茜莖……1 棵 #剁碎

做法 @ 醬汁混合後，倒入雲耳內拌勻即可享用！

comments...

薄荷薯仔

Gigi Wong 夏日消暑沙律 ☀ 薄荷透心涼 ⛄⛄ 好易整 ✓

將薯仔和調味料放入密實袋內搖勻，味道會更加均勻。

材料 @ 新薯 .... 8 個 # 新薯加入到面水，放入微波爐「叮」3 分鐘，
瀝去水分

調味料 @ 薄荷葉 .... 2 棵份量 # 切碎
鹽 .... 1 茶匙
意大利陳醋 .... 1 湯匙
初榨橄欖油 .... 2 湯匙

做法 @ 薯仔不用去皮，切大塊。
@ 調味料拌勻後，趁熱加入已焓熟薯仔，會更易入味，拌勻即可。

cold dish_snack

涼拌腐竹

Gigi Wong 清爽涼菜 豆香濃⊙好易整 ✓

comments...

材料 @ 腐竹......3 枝 #浸軟，切度，汆水 5 分鐘
木耳絲......15 克 #浸軟，汆水
大豆芽菜......50 克 #洗淨
薑......2 片 #切絲
紅西椒......半個 #切絲

調味料 @ 鹽......1 茶匙
麻油......1 茶匙

做法 @ 熱鑊下油，下薑絲，爆炒大豆芽菜，加入木耳絲、腐竹和紅西椒絲，最後加入調味，拌勻，涼後即可享用。

comments...

cold dish_snack

香酥葱油餅

Gigi Wong 簡易點心 ✓ 卜卜脆 ☺☺ 趁熱食 ♨

comments...

圓形餃子皮

#肥肉內加鹽，要剁至這樣糜爛才算合格。

材料 @ 圓形餃子皮 ....半斤

餡料 @ 葱粒............1 杯
肥豬肉 ....1 塊 # 剁至極幼
鹽............半茶匙 # 加入肥豬肉內同剁

做法 @ 餃子皮上鋪一層薄薄的肥肉末，灑上葱粒，蓋上餃子皮，再鋪一層薄薄的肥肉末和葱粒，蓋上第三片餃子皮。
@ 將餃子皮邊緣捏實。
@ 平底鑊內下少許油，用慢火煎至金黃即可享用。

Flaky spring onion pancakes

cold dish_snack

蒜泥白肉

Gigi Wong 四川風味涼菜 ❋ 麻辣香♨♨ 好開胃 ✓

這是蒜泥白肉的簡化版，將清新的青瓜、肥瘦兼備的白肉、麻香的醬汁混在一起，吃時別有一番風味。

材料 @ 五花腩 .... 1 塊約半磅
青瓜 .... 1 條 # 用刨刨成長條

料頭 @ 薑 .... 2 片
葱 .... 1 條
紹興酒 .... 2 湯匙

醬汁 @ 葱白粒 .... 2 湯匙 # 用石臼舂成茸
薑粒 .... 2 湯匙 # 用石臼舂成茸
蒜茸 .... 2 湯匙 # 用石臼舂成茸
生抽 .... 1.5 湯匙
鎮江香醋 .... 1 湯匙
糖 .... 1 湯匙
麻油 .... 1 茶匙
花椒粉 .... 1 茶匙
辣椒油 .... 1 湯匙

做法 @ 把五花腩與料頭同放入冷水中煮滾，約 15 分鐘，熄火焗 5 分鐘，共做兩次；取出浸冰水，再放入雪櫃半小時。
@ 把已刨成長條的青瓜排在碟上。
@ 把五花腩切成薄片放在青瓜條上。
@ 淋上已拌勻的醬汁即成。

comments...

把五花腩浸冰水，再放入雪櫃半小時，可令五花腩肉質硬點，容易切成薄片。

comments...

米飯
麵食

rice_noodles

雞酒米粉

Gigi Wong 酒香撲鼻 ♥ 微微藥材香 ✽ 感覺很滋補 ♨

comments...

材料 @ 米粉 .... 1 份 # 用水略浸
雞上髀肉 .... 2 塊 # 切塊，略醃
薑 .... 6 片
麻油 .... 2 湯匙
片糖 .... 1/8 片
花雕酒 .... 半杯
杞子 .... 1 湯匙
當歸 .... 1 片
黃芪 .... 2 片
蒜茸 .... 2 湯匙
滾水 .... 2 杯

醃雞料 @ 紹酒 .... 1 湯匙
糖 .... 半茶匙
胡椒粉 .... 適量

做法 @ 把米粉煮熟，撈起放入大碗內

@ 用麻油起鑊爆炒薑片，至香味四溢，倒入雞肉兜炒，把花雕酒倒入煮滾，放入其餘材料，冚蓋煮 5 分鐘。

@ 如能接受，可於雞酒起鍋前再加 2 湯匙花雕酒，倒入已煮好的米粉內即可享用。

米飯
麵食

rice_noodles

小米飯

Gigi Wong 健康米飯 ♥ 顏色金黃 ✽ 味清香 ♨

小米能補虛損、益丹田，宜體質羸弱、胃納欠佳者享用。小米的質地較硬，如用來蒸飯或煲粥一定要浸過夜，蒸煮後的質感才會軟綿。

材料 @ 小米……1 杯
油……1 湯匙
杞子……適量 # 沖淨

做法 @ 小米浸過夜，洗淨，瀝去水分，把油倒在小米上，拌勻，讓每粒小米都沾上油。
@ 蒸籠鋪上紗布，倒入小米，猛火蒸 30 分鐘即可，享用前加入已清洗的杞子。

未煮前的小米

米飯
麵食

rice_noodles

菜脯炒飯

Gigi Wong 愛惜食材 ♥ 不浪費 ✽ 好美味 ♨

如將每餐剩下的飯倒掉實在非常浪費，我將剩飯放入冰格貯存，儲夠一定份量就用來炒飯，既美味又環保。

材料 @ 雞蛋……2 個 # 打勻
剩飯……4 碗
菜脯粒……2 湯匙
臘腸……2 條 # 切粒
臘肉……1/3 條 # 切粒
芥蘭梗……2 棵 # 切粒
雲耳……幾朵 # 浸軟後切碎
冬菇……2 朵 # 浸軟後切粒
薑米……2 湯匙
葱花……1 杯

做法 @ 燒熱鑊，下 1 湯匙油，加入臘腸、臘肉粒，用中火兜炒至出油，加入菜脯粒、芥蘭梗、雲耳、冬菇和薑米，兜炒約 2 分鐘盛起。
@ 原鑊下少許油，倒入蛋液，在蛋液未凝固前，加入剩飯，待飯炒鬆了，倒入各粒料，炒勻後加入葱花，兜勻後即可上碟。

comments...

小菜 main dish

鮮蟲草花炒牛肉

Gigi Wong 有益菇菌 ♥ 滋肺補腎 ✽ 抗衰老 ♨

comments...

材料 @ 牛肉……225 克 # 切片，略醃
鮮蟲草花……120 克 # 洗淨，剪去根
洋葱……半個 # 切絲

料頭 @ 乾葱頭……2 粒 # 切碎
薑……4 片
葱白……2 條 # 切度

牛肉醃料 @ 糖……1/4 茶匙
鹽……半茶匙
胡椒粉……半茶匙
生粉水……1 茶匙 # 生粉 1 茶匙、水 1 湯匙調勻
油……1 茶匙

調味料 @ 鹽……1 茶匙
蠔油……1 湯匙

做法 @ 鑊內下 1 湯匙油爆香料頭，倒入牛肉兜炒 10 秒即盛起。
@ 原鑊再下少許油爆炒洋葱及鮮蟲草花，牛肉回鑊，下調味料兜勻即可。

乾蟲草花

新鮮蟲草花

牛肉炒至僅熟即可，因要回鑊再兜炒。

這菜式用乾、鮮蟲草花皆宜 # 乾蟲草花烹調前要浸開才用 # 鮮蟲草花的口感較爽，而乾蟲草花則較韌。

小菜 main dish

全家福

Gigi Wong 美味燜菜 ♥ 墨魚味鮮 ✽ 拌飯一流 ♨

如你親自剁肉做丸仔，這菜式可以說是複雜的；如你買現成的肉丸，這可以說是懶人菜。全家福的份量頗多，煮一大鍋，隔天吃味道更佳。

材料	@	雞	1 隻 # 斬件
		排骨	1 大塊 # 斬件
		急凍豬肚	1 個 # 洗淨，切塊
		墨魚	1 條 # 去衣，切塊
		雞蛋	4 個 # 烚熟，去殼
		麵粉	少許
		酒	少許
		紅棗	8 粒 # 去核
醃肉底味	@	鹽	1 茶匙
		糖	半茶匙
		生粉	1 湯匙
		胡椒粉	1 茶匙
		麻油	1 茶匙
肉丸料	@	冬菇	2 朵 # 切碎
		免治肉	半斤 # 上底味
		蛋液	1 個份量
		麵粉	適量
料頭	@	蒜頭	4 粒
		薑	2 片
		指天椒	1 隻
		冰糖	2 粒
		八角	1 粒
調味料	@	生抽	1 湯匙
		老抽	1 湯匙
		蠔油	1 湯匙
		鹽	1 茶匙

做法 @ 雞、排骨、豬肚和墨魚一起上底味，拌勻。

@ 把冬菇碎加入免治肉內拌勻，做成肉丸，沾蛋液再上麵粉，放入油鑊中炸至金黃撈起。

@ 雞蛋上薄麵粉，放入油鑊中略炸至金黃撈起。

@ 熱鑊冷油爆香料頭，加入雞件、排骨、豬肚和墨魚，灒酒，下調味，炒勻，注過面滾水，放入紅棗，收慢火同燜半小時。

@ 加入肉丸和雞蛋，同燜至汁液濃稠即可。

雞蛋撲上薄麵粉再油炸，除了顏色金黃漂亮外，也不易燜爛。

Chinese surf-and-turf stew

肉丸可買現成的，能省卻一半烹調工夫。

小菜 main dish

生炒金沙骨

Gigi Wong 懶人首選菜式 ♥ 好易整 ✽ 開胃菜 ♨

comments...

材料 @ 金沙骨........................1 斤 # 切塊

調味料 @ 冰糖........................2 粒
魚露........................2 湯匙
水........................半杯 # 視情況逐少加

做法 @ 用少許油爆炒金沙骨至肉變白色。
@ 加入冰糖，炒至肉呈微黃色，加入魚露、水，蓋上鍋蓋燜至汁液濃稠即可。

加入冰糖，炒至微黃的金沙骨

Pan-seared spareribs

comments...

小菜 main dish

紅麴豬手

Gigi Wong 天然健康色素 ♥ 皮軟肉腍 ✿ 豐富骨膠原 ♨

蒸了 1.5 小時的豬手，口感腍滑；如想爽口些，蒸 1 小時便可。
待豬手蒸了 45 分鐘後反轉，可以令顏色更均勻。

材料 @ 豬手 .... 1隻 # 斬件，上底味
紅麴米、水 各2湯匙
糯米粉 1湯匙
有衣花生 .... 半杯 # 浸半小時

豬手底味 @ 鹽 1茶匙
生抽 1湯匙
胡椒粉 .... 少許
紹酒 2湯匙
紅糖 2茶匙

做法 @ 把已上了底味的豬手，混合其餘材料拌勻，先蒸45分鐘。
@ 反轉豬手再蒸45分鐘即可。

紅麴米

Steamed pork trotter in red yeast rice paste

comments...

小菜 main dish

鄉里炒粒粒

Gigi Wong 鄉土菜 ♥ 廚餘再用 ✽ 愛惜食材 ♨

豬油渣可以用肥肉代替豬膏，將它切粒，炸香即可。

以前的鄉下人愛惜食物，物盡其用，會將蘿蔔梗收集起來做菜 # 現在的白蘿蔔梗在推出市場前已棄掉，故用芥蘭莖代替。

材料切成粒狀，調味香濃惹味，舀一大匙，就可以拌一大碗飯，切合以前農村的民情。

材料 @ 免治豬肉……200 克 # 上底味
豬油渣……2 湯匙 # 切碎
芥蘭莖……1 杯 # 切粒
豆豉……2 湯匙 # 用溫水浸洗
紅椒……1 隻 # 切粒
皮蛋……1 個 # 切粒
蒜頭……2 粒
薑……2 片

底味 @ 鹽……半茶匙
胡椒粉……少許

調味料 @ 糖……1 茶匙
蠔油……1 湯匙

做法 @ 用少許油爆香蒜頭，下免治豬肉兜炒至略熟，盛起待用。
@ 原鑊用剩油爆香薑片，倒入芥蘭莖粒炒勻，加入豆豉、紅椒粒，把免治豬肉倒入同炒勻，最後加豬油渣及皮蛋粒，下調味料兜勻即可上碟。

comments...

小菜　main dish

鹹酸菜燜豬大腸

Gigi Wong 香濃惹味菜 ♥ 鹹香酸辣 ✽ 翻煮更美味 ♨

鹹酸菜買回家後，我會處理過才用來烹調菜餚，經過處理後的鹹酸菜，味道會更香濃。

材料 @ 急凍豬大腸……2 條約 2 斤 # 汆煮 1 小時
鹹酸菜……1 棵 # 切片，處理法看做法第一點

煮大腸料 @ 白胡椒粒……1 湯匙
薑……2 片
大蒜……1 棵
米酒……2 湯匙

料頭 @ 薑……6 片
葱……4 條 # 切度
芫茜……2 條 # 切度
乾紅椒……1 隻 # 切斜刀

香料 @ 草果……2 粒
甘草……2 片
香葉……2 片
八角……2 粒
陳皮……1/3 塊
花椒粒……2 茶匙 # 用茶包袋裝好

調味料 @ 豆瓣醬……2 湯匙
蠔油……2 湯匙
生抽……1 湯匙
白醋……1 湯匙
糖……1 茶匙
胡椒粉……1 茶匙
滾水……1 杯

做法 @ 鹹酸菜處理法：先摘小塊鹹酸菜試味，若太鹹，應先用鹽水浸半小時，可把鹹味釋放。鹹酸菜切片後，放入白鑊內烘乾，烘乾後改用細火，然後加 2 茶匙糖及 1 茶匙生油煨一煨就可採用。

@ 切去大腸肥膏；大腸和煮大腸料放入鍋內，加入過面水，用中火煮 1 小時，取出切斜片。

@ 熱鑊冷油順序爆炒料頭，收慢火，再把香料加入，炒出香味後（可棄去花椒袋）放入大腸及鹹酸菜，再把調味料加入燜 10 分鐘即可！

comments...

香料包：草果、甘草、香葉、八角、陳皮、花椒粒。

在處理步驟中，改用細火後才可加糖進鹹酸菜內，因為糖容易「搶火」，會令鹹酸菜焦燶；加油的作用是讓鹹酸菜有油的滋潤。

小菜 main dish

燉元蹄

Gigi Wong 豬皮肥肉最美味 ♥ 軟腴香滑 ✽ 不怕胖 ♨

如沒有壓力鍋，可待元蹄上色後，先用大火煲滾，再改為慢火煲 2.5 小時。

材料 @ 元蹄……1 個約 2 斤 # 汆水
青菜……1 斤 # 洗淨
薑……4 片
葱……4 棵 # 打結
紅棗……8 粒 # 去核

調味料 @ 紹酒……2 湯匙
老抽……1 湯匙
生抽……2 湯匙
冰糖……3 粒 # 舂碎
紅麴米……1 湯匙 # 用茶包袋裝好

焯菜料 @ 油……1 湯匙
鹽……1 茶匙
水……2 杯

做法 @ 熱鑊下油，把已索乾水的元蹄煎至四面金黃，取出備用。
@ 倒出多餘油，放入薑、葱爆香，再順序把調味料加入，放入紅棗，注入過面滾水，放入元蹄不停滾動至上色。
@ 將元蹄轉到壓力鍋煲半小時，取出看尚餘多少水分，用大火收汁，途中以汁淋在元蹄上，以上色入味。
@ 煮滾焯菜料，把菜倒入，冚蓋，2 分鐘即可取出，排在碟上，把元蹄放在中央，澆上濃汁即可。

小菜 main dish

蜜汁豬頸肉

Gigi Wong 肉爽脆 ♥ 蜜味濃 ✿ 好味道 ♨

蜂蜜不能受高溫，故此要確定肉已熟，才倒入蜜汁，而且只能用中火。
照燒汁是日式調味料，亦可以用柱侯醬代替。

Gigi Wong #煎豬頸肉時，要用慢火細煎，並要有耐性，煎至有點焦香，才夠香口。

材料 @ 豬頸肉 4塊 # 用醃料醃過夜

醃料 @ 照燒汁 210毫升
紹興酒 半杯
鹽 1茶匙

蜜汁料 @ 蜂蜜 3湯匙
黑椒汁 2湯匙
喼汁 1湯匙

做法 @ 平底鑊內下少許油，用慢火把肉煎至兩面略帶焦黃，待肉全熟即可倒入蜜汁，煮至略收乾即可。
@ 也可以用烤代替煎。用焗爐上火230℃，每邊燒烤10分鐘，取出搽上蜜汁，轉用180℃每邊再燒2分鐘即可。

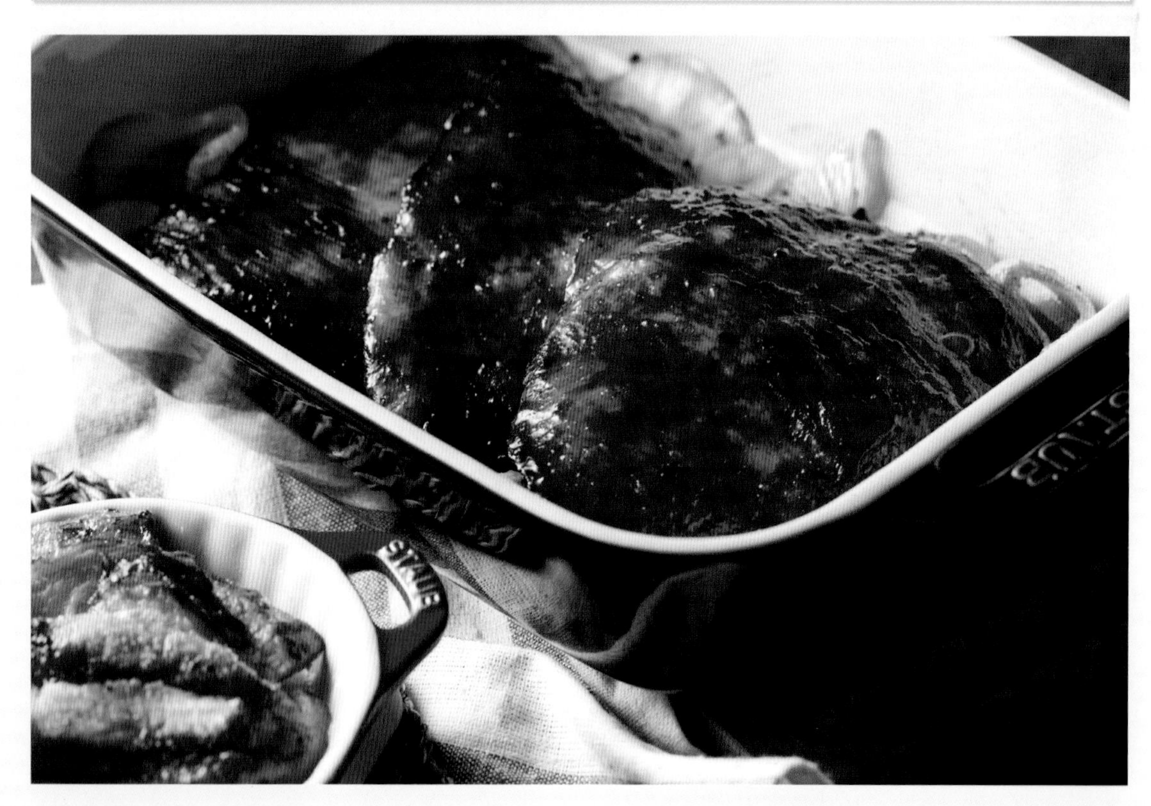

小菜 main dish

沙葛炒五花腩

Gigi Wong 夏日開胃菜 ♥ 沙葛好爽脆 ✽ 好易煮 ♨

有鹹酸菜的菜式，是伴飯的首選。五花腩宜薄切，因吸味容易，故在烹調前，宜將五花腩冷藏至硬，這樣才容易切成薄片。

材料 @ 沙葛 350克約半個 # 撕去皮，切絲，氽水
鹹酸菜 半個 # 切絲，浸水15分鐘
五花腩 200克 # 去皮切片，上底味
韭菜花 90克 # 切度
指天椒 1隻 # 切絲
薑 2片 # 切絲
蒜頭 2粒 # 切幼粒
紹興酒 1湯匙

五花腩底味 @ 鹽 半茶匙
胡椒粉 少許

調味料 @ 鹽 半茶匙
糖 半茶匙
生粉 1茶匙
蠔油 1湯匙
水 1湯匙
胡椒粉 少許

做法 @ 白鑊烘乾鹹酸菜，約5分鐘，改慢火，加2茶匙糖，兜勻後加1茶匙油，讓其油潤，即可盛起，待用。
@ 熱鑊下油，爆香薑、蒜後加入肉片，兜炒，待肉片散開後，濳酒，加入沙葛、鹹酸菜，兜炒一會，下調味料，最後加入韭菜花、指天椒絲，略兜炒一會即可上碟！

Stir fried pork belly with yam bean

沙葛 # 去皮後顏色雪白 # 質地爽脆、味道清甜，可作沙律享用。

將沙葛去皮是有竅門的，先在頂部輕剔一刀，再在開口處撕去沙葛皮。

小菜 main dish

剁椒蒸排骨

Gigi Wong 別被紅噹噹騙倒 ♥ 只有小小辣味 ✽ 芋頭索汁好好味 ♨

鋪上番茄，除了令顏色美麗外，也有增鮮去辣味的功效。

材料 @ 排骨 450克 # 切塊，用鹽2茶匙抓洗兩次，沖水後索乾，加醃料備用

芋頭 半個 # 去皮，切片

番茄 半個 # 切細粒

排骨醃料 @ 鹽 1茶匙

糖 2茶匙

麻油 1茶匙

老抽 1茶匙

蠔油 1湯匙

豆豉 1湯匙

剁椒 1湯匙

粟粉 2茶匙

做法 @ 芋頭鋪平在碟底，放上已醃好的排骨。

@ 再把番茄鋪上，猛火蒸12分鐘即可。

凡是蒸肉類、海鮮必須用大火，可迅速收緊外皮，鎖着肉汁，保留原汁原味。

Steamed pork ribs with pickled chopped chillies

comments...

小菜 main dish

紅煨牛肉

Gigi Wong 牛脹軟腍 ♥ 微辣 ✲ 惹味開胃菜 ♨

comments...

Gigi Wong # 以前我會將汆牛肉的湯水倒掉，但後來發覺汆了牛肉半小時的湯水，撇去浮油，更有濃濃的牛肉鮮味，所以現在會將它留用。

紅辣椒是這菜式的點睛之處，鮮香中帶點微辣，越吃越開胃。

Braised beef shin in
aromatic soybean paste

comments...

材料 @ 牛脲……500 克 # 凍水加入薑 2 片、牛脲，汆煮半小時，撇去浮泡，湯汁留用

京葱……2 棵 # 斜刀切度

料頭 @ 冰糖……2 粒

薑……2 片

蒜頭……2 粒

八角……1 粒

紅辣椒……1 隻

調味料 @ 原磨豉醬……1 湯匙

老抽……1 湯匙

紹興酒……2 湯匙

鹽……半茶匙

做法 @ 將牛脲切厚塊，待用。

@ 用小火下少許油，將料頭爆香，下調味料，放入牛脲，注入煲牛脲原湯至肉面，用中火燜 1 小時至腍。

@ 用少許油爆香京葱，盛起圍在碟邊，中央放牛脲，即可享用。

comments...

小菜 main dish

洋葱鴨

Gigi Wong 洋葱已糖化 ♥ 鴨肉嫩滑 ✽ 拌飯一流 ♨

調味料的份量是相當易記的，由 1 數到 5，1 杯水、2 湯匙糖……5 湯匙老抽。

材料 @ 鴨 .... 1 隻 # 洗淨，索乾水分
洋葱 .... 8 個 # 切塊

調味料 @ 老抽 .... 5 湯匙
生抽 .... 4 湯匙
紹酒 .... 3 湯匙
糖 .... 2 湯匙
水 .... 1 杯

做法 @ 以 3 湯匙老抽抹勻鴨身，放入油內煎至鴨皮金黃色，備用。
@ 把部分洋葱塞入鴨肚，剩下的洋葱鋪上鍋內，鴨放其上，淋上調味，燜 60 至 90 分鐘至腍即成。

comments...

Braised duck in onion sauce

經過個多小時的燜煮，洋葱已糖化，是非常美味的。

在圖中，我看似有點暴力，拿着鴨頸在鑊上翻來覆去將鴨皮煎至金黃色，但這樣卻是最方便、兼容易處理。各位煮持人，你們也試試吧！

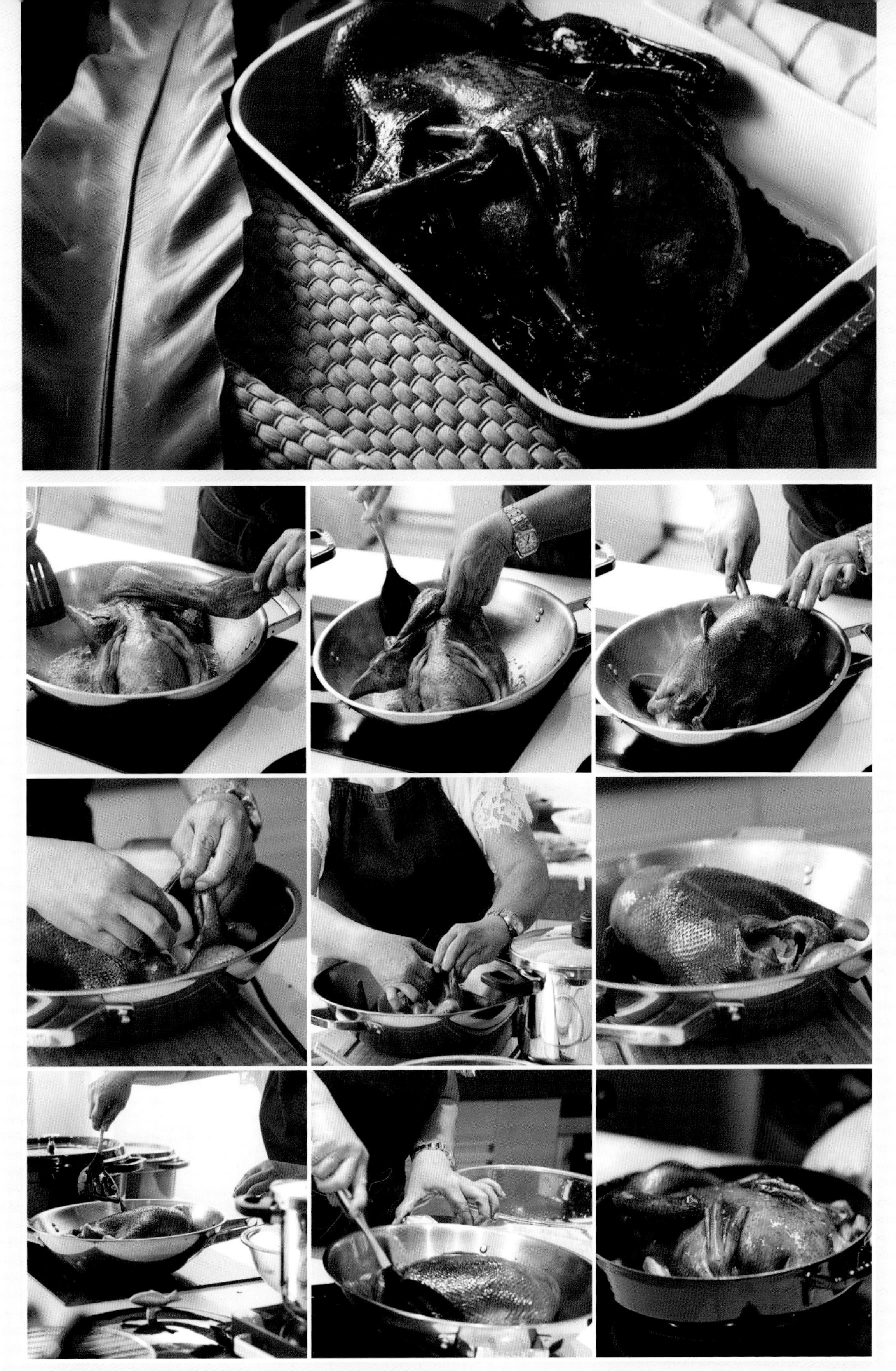

小菜 main dish

冬瓜燜鴨

Gigi Wong # 宜請檔主挑選嫩鴨炮製這菜式，嫩鴨的肉味雖然較「削」，但肉質軟嫩，燜及紅燒皆美味可口。

comments...

材料 @ 鴨 .... 1隻 # 斬件，上底味
冬瓜 .... 1斤 # 去瓤，切大塊
薑 .... 1塊 # 拍扁
蒜頭 4粒

底味 @ 鹽 1茶匙
胡椒粉 .... 少許

配料 @ 芡實 .... 2兩 # 洗淨，浸1小時
薏米 .... 2兩 # 洗淨，浸1小時
陳皮 .... 1角 # 浸軟，刮去瓤

調味料 @ 蠔油 1湯匙
鹽 1茶匙

做法 @ 熱鑊冷油爆香薑塊、蒜頭，放入鴨件和配料，加入過面滾水，轉用中火焗45分鐘。
@ 倒入冬瓜繼續焗半小時，加入調味即可享用。

 # 此菜式不能加生抽，否則味道會變酸。

comments...

小菜 main dish

青瓜炒雞柳

Gigi Wong 夏日時令菜 ♥ 青瓜清甜爽 ✿ 簡易晚餐 ♨

市面有不同品種的青瓜出售，宜選用有刺青瓜；經過處理的青瓜會更爽口。

Gigi Wong # 嗜辣者，可以用指天椒代替甜椒。
Gigi Wong # 可以用豬頸肉、蝦仁或蝦乾代替雞肉，均有不同的風味。

comments...

材料 @ 青瓜……2條 # 切厚片
雞上髀肉……1塊 # 切粗條，上底味
紅甜椒……1個 # 切粗條
葱白……1條 # 切度
蒜頭……2粒

底味 @ 鹽……1茶匙
黑椒粉……1茶匙

調味料 @ 蠔油……1湯匙
糖……半茶匙
麻油……1茶匙

做法 @ 把青瓜片放入膠袋內，加入1茶匙鹽，揑緊袋口，用力搖使青瓜出水，沖凍水後揸乾。

@ 熱鑊下油1湯匙，爆香蒜頭，把雞肉爆炒至變白，盛起。原鑊加入青瓜，兜勻，下雞肉，隨後加紅甜椒及葱白，略兜炒即可下調味，兜勻上碟。

小菜 main dish

★☆☆☆

大盤雞

Gigi Wong 四川風燜雞 ♥ 濃濃花椒味 ✽ 好惹味 ♨

comments...

材料 @ 雞……1隻 # 斬件，加底味
花椒……1湯匙
薯仔……4個 # 去皮，切塊
啤酒……1小枝
青紅椒……各1隻 # 切塊

底味 @ 鹽……1茶匙
胡椒粉……少許

料頭 @ 冰糖……2粒
薑……4片
蒜頭……4粒
葱……4條 # 切度
草果……1粒
八角……1粒
桂皮……1片
香葉……2片

調味料 @ 老抽……2湯匙
生抽……1湯匙
鹽……2茶匙

做法 @ 油鑊中放入花椒，炒至花椒香味溢出，撈起花椒，棄掉。
@ 原鑊再放入料頭，爆炒至香味溢出，倒入雞件，炒至雞肉變色（呈白色），然後倒入薯仔，注入啤酒，燜約15分鐘。
@ 15分鐘後或至汁液濃稠，加入青紅椒，兜勻，倒入調味料炒勻即可。

Gigi Wong # 忙裏偷閑，趁着燜餸的空檔，打個電話傾下計先。

comments...

香料是這菜的靈魂，令雞肉的味道變得豐富，尤其現今多採用冰鮮雞，香料兼可以辟去雪味。

Beer braised chicken with Sichuan peppercorns and spices

comments...

小菜 main dish

三文魚炒大豆芽

Gigi Wong 魚油豐富 ♥ 含 Omega-3 ✽ 有益身體 ♨

三文魚魚油豐富，烹調時只須加入少許油兜炒便可。

材料 @ 三文魚……1 塊約 200 克 # 連皮切粗條，上底味
薑……4 片 # 切絲
大豆芽菜……160 克
鮮冬菇……4 朵 # 切絲
黑木耳……20 克 # 浸軟，切絲
木魚片……1 包

底味 @ 鹽……半茶匙
胡椒粉……少許

調味料 @ 鹽……半茶匙
米酒……1 湯匙

做法 @ 鑊內下少許油，用中火把三文魚炒至出油，盛起備用。
@ 用鑊內剩下的油爆香薑絲，倒入大豆芽菜、冬菇絲、木耳絲，炒勻，灒酒，下鹽調味，三文魚回鑊，炒香盛起，灑上木魚片即可。

comments...

comments...

魚香蝦球

豆瓣醬以四川郫縣出產的為佳，當地盛產蠶豆，氣候濕潤，全年平均氣溫約在攝氏十七度，有利做醬。

Gigi Wong #豆瓣醬是川菜的主要調味料之一，有「川菜之魂」的稱號
#嗜辣者，不妨找郫縣出產的豆瓣醬試試。

comments...

#魚香汁較濃稠，也容易掛在蝦球面。

Fried prawns in Sichuan Yu Xiang sauce

comments...

材料 @ 中蝦……10隻 # 去殼，用鹽抓洗2次，索乾，蝦背𠝹一刀，挑腸
生粉……適量
豆瓣醬……1湯匙
紹興酒……2湯匙

料頭 @ 指天椒……2隻 # 切碎
葱白……2條 # 切碎
薑米……4片
蒜茸……4粒份量

魚香汁 @ 生粉……1湯匙
生抽……2湯匙
糖……3湯匙
山西老陳醋……4湯匙

做法 @ 蝦撲上少許生粉，煎至七成熟，盛起。
@ 爆香料頭，加入豆瓣醬，略炒，灒酒，轉調大火，注入魚香汁，煮滾後加入蝦，兜勻即可上碟。

小菜 main dish

★☆☆☆

鹹檸檬醬蒸魚

Gigi Wong 夏日醒胃菜 ♥ 魚肉鮮甜 ✽ 蒸魚無難度 ♨

comments...

材料 @ 烏頭......1 條約斤半 # 去鰓、去鱗，劏開肚，洗乾淨
葱......2 棵 # 切半
指天椒......1 隻 # 切絲

檸檬醬 @ 酸薑......4 片 # 切絲
鹹檸檬......1 個 # 切絲
酸梅......3 粒 # 揉爛，去核
酸梅醬......4 湯匙
蕎頭......8 粒 # 切絲
糖......1 湯匙
水......半杯

做法 @ 碟內鋪上葱條，將烏頭從魚腹掰開放在葱條上。
@ 猛火蒸 6 分鐘至熟，取走葱條。
@ 將檸檬醬料煮出味後，加入紅椒絲，淋在魚上即可。

comments...

Gigi Wong

comments...

Steamed grey mullet
in salted lemon paste

comments...

Gigi Wong # 煮餸前，預先切料頭，例如：辣椒粒、蒜茸、葱粒等等，需要用時就不會手忙腳亂再張羅。

comments...

小菜 main dish

肉桂雞髀菇

Gigi Wong 有益菇菌 ♥ 肉桂味香濃 ✽ 烹調簡易 ♨

comments...

材料 @ 雞髀菇……500克 # 抹乾淨，切滾刀塊
蒜頭……4粒 # 切碎
紅辣椒……1隻 # 切碎
葱……2條 # 切碎
芫茜……1棵 # 切碎

調味料 @ 水……1杯
紅糖……2湯匙
白糖、老抽……各1湯匙
肉桂粉……1茶匙

做法 @ 白鑊炒雞髀菇約3分鐘，以去水分，盛起。
@ 用1湯匙油爆香蒜茸及紅辣椒碎，順序加入調味，煮至濃稠，把雞髀菇倒入，兜勻，再把葱花、芫茜加入炒勻即可。

小菜 main dish

三杯菇

Gigi Wong 三杯雞變奏版 ♥ 九層塔是亮點 ✿ 好味道 ♨

凡烹調鮮菇前，用濕布抹乾淨就可，不要用水洗，否則鮮菇如海綿般吸飽水分，烹調時菜式會水汪汪。

材料 @ 雞髀菇……4個 # 切塊
鮮冬菇……4朵 # 一開二
本菇……1包
麻油……1湯匙
菜油……1茶匙
九層塔葉……1杯 # 後下

料頭 @ 洋葱……半個 # 切絲
薑……2片
蒜頭……2粒
紅辣椒……1隻

調味料 @ 黑醋……1湯匙
紅糖……1茶匙

做法 @ 用菜油、麻油爆香料頭，倒入三菇，爆炒幾分鐘。
@ 加入調味料，炒勻後倒入九層塔葉，兜炒均勻即可享用。

剪去菇蒂，吃時有更軟滑的口感。# 菇類可按時令購買。

因麻油易搶火，故加入菜油一同爆炒三菇，既能保留麻油的香味，也不易焦燶。

小菜 main dish

韭菜鮮蝦

Gigi Wong 夏日醒神菜 ♥ 鮮甜撞大辣 ✽ 味蕾新激盪 ♨

烹調這道菜最緊要快，因為過熟的韭菜會韌。

Stir fried fresh prawns with Chinese chives

comments...

這菜式最理想是用河蝦，但河蝦不易買到，故用海蝦或基圍蝦代替；配上味濃的香草、蔬菜，香氣撲鼻。

材料 @ 鮮蝦..........200 克 # 剪去鬚、腳
韭菜..........100 克 # 洗淨切度
指天椒..........2 隻 # 切碎
紫蘇葉..........30 克 # 洗淨，切碎

調味料 @ 糖..........半茶匙
鹽..........1 茶匙
麻油..........1 茶匙
胡椒粉..........少許

做法 @ 用略多油爆炒鮮蝦，炒至變色即可盛起，瀝去多餘的油。
@ 用原鑊煸香指天椒，放入鮮蝦、韭菜及紫蘇葉，加調味料，兜勻即可上碟。

comments...

小菜 main dish

黃芽白包釀蝦膠

Gigi Wong 宴客菜 ♥ 蝦膠彈牙 ✽ 清甜美味 ♨

如沒有黃芽白，可用椰菜代替，但黃芽白比較清甜。

材料 @ 蝦仁……半斤 # 用鹽抓洗兩次，索乾水分，拍成膠狀
肥肉……2 兩 # 剁至極幼
黃芽白……12 片 # 只要葉部，汆水 2 分鐘
蛋白……1.5 個 # 打匀

蝦仁調味 @ 胡椒粉……半茶匙
蛋白……半個
麻油……半茶匙
紹興酒……1 茶匙

芡汁 @ 生粉……1 茶匙
水……1 湯匙
麻油……1 茶匙
鹽……半茶匙
蒸菜汁……半杯

做法 @ 蝦膠用調味醃半小時，加入肥肉，順一方向攪匀，撻打幾下。
@ 把汆過水的黃芽白抹乾，鋪平在砧板上，放上少許蝦膠包成荷包狀；排列在碟上。
@ 猛火隔水蒸 5 分鐘，取出後潷出菜汁。菜汁與其餘芡汁料拌匀，煮成薄芡，熄火，把蛋白倒入，拌匀，淋上荷包面即可享用。

Gigi Wong # 將蝦肉順一方向攪勻，會較容易起膠。

comments...

小菜 main dish

啫啫鯇魚塊

Gigi Wong 鯇魚腩嫩滑 ♥ 欖角好惹味 ✽ 啖啖魚鮮香 ♨

用砂鍋烹調這菜式有好多優點，除了能保溫外，還可以原鍋上桌，保留原汁原味。

材料 @ 鯇魚……1 件 # 切粗條，用醃料醃 1 小時
紹興酒……1 湯匙

醃料 @ 鹽……1 茶匙
胡椒粉……少許
生粉……1 茶匙
欖角……4 粒 # 剁碎，加糖 1 茶匙蒸 10 分鐘
麻油……1 茶匙

料頭 @ 薑……8 片
南薑……8 片
蒜頭……8 粒
紅葱頭……8 粒
葱白……2 條 # 切度

做法 @ 砂鍋內下 2 湯匙油爆香料頭，待香氣四溢時，把魚塊平鋪在配料上。
@ 冚蓋，在蓋邊下酒，焗煮 6 分鐘即可享用。

---

comments...

在砂鍋蓋邊下紹興酒，讓魚肉均勻地有酒香。

Sizzling grass carp fillet in clay pot

comments...

小菜　main dish

麻醬蒜泥蒸茄子

Gigi Wong 簡單易做 ♥ 芝麻香撲鼻 ✽ 凍食也美味 ♨

comments...

材料 @ 茹子 .... 2 條 # 切度，剘花
麵粉 .... 適量

調味料 @ 芝麻醬 .... 2 湯匙
鎮江香醋 .... 2 湯匙
蒜泥 .... 1 粒份量
鹽 .... 半茶匙
糖 .... 1 茶匙

做法 @ 茹子輕輕灑上麵粉，放碟內蒸 7 分鐘。
@ 取出，澆上已調勻的調味料即可享用。

comments...

這些調味料看似是平凡不過，但每樣一點，調勻後就非常美味。

Steamed eggplant with sesame paste and garlic

comments...

湯 soup

秋天滋潤湯

Gigi Wong 秋天時令湯 ♥ 滋潤解燥 ✽ 有益身體 ♨

comments...

材料 @ 海龍、無花果..........各 1 兩 # 略沖洗
沙參、玉竹..........各 1 兩 # 略沖洗
百合、杏仁..........各 1 兩 # 略沖洗
淮山..........1 兩 # 略沖洗
黑豆..........1 兩 # 浸半小時
乾螺頭..........4 兩 # 處理好
陳皮..........1 個 # 浸軟，刮去瓤
紅蘿蔔..........1 條 # 去皮，切塊
水..........18 杯
鹽..........適量 # 後下

做法 @ 乾螺頭處理法：1 斤乾螺頭沖洗乾淨後，用過面水浸一夜，連水倒入電飯煲，加入 1 粒冰糖，按下煲掣，直到水乾即可。攤涼後分別裝在三文治袋，放入冰格，用時取出，既方便，又省時。

@ 將所有材料放入鍋內，先用猛火煲滾，10 分鐘後改用中慢火煲 3 小時即可。

@ 飲用前先試味才下鹽。

海龍

Nourishing herbal soup for autumn

comments...

花生木瓜魚仔湯

Gigi Wong 養顏健體湯 ♥ 魚味濃郁 ✽ 湯鮮甜 ♨

這湯成本低，烹調簡易；功效強身健體，一家老幼皆宜飲用。

Gigi Wong # 紅棗是煲老火湯常用的湯料，有養顏防老、補氣養血、健脾胃、潤心肺、防掉髮、強健筋骨等作用。

comments...

材料 @ 魚仔……1 斤 # 煎香放入魚袋
花生……1 杯 # 連衣沖洗
熟木瓜……1 個 # 去皮，去籽，切塊
腰果……1/3 杯 # 沖淨
栗子……1/3 杯 # 沖淨
紅棗……15 粒 # 拍扁，去核
陳皮……1 個 # 浸軟，去瓤
水……18 杯
鹽……適量 # 後下

做法 @ 所有材料（除了鹽）放入鍋內，先用大火煲滾 10 分鐘後，改用中慢火煲 3 小時。
@ 飲用前先試味才下鹽。

comments...

Gigi Wong # 用魚袋盛着魚仔煲湯，以免喝湯時被魚骨刺傷。

comments...

花生衣有補血作用，千萬別丟掉。做化療後用花生衣煲水飲，能補充血液氧分。

Fish soup with peanuts and papaya

食譜內用的魚仔是油鰔仔，可用魚尾代替魚仔。

湯 soup

菜乾蘋果湯

Gigi Wong 滋潤湯 ♥ 濃濃菜乾香 ✽ 響螺好滋陰 ♨

comments ...

材料 @

不見天	1 斤	# 汆水
乾螺頭	4 兩	# 處理好，處理方法看 P.89
白菜乾	4 兩	# 洗淨浸軟，去沙
蘋果	4 個	# 一開二，去芯
雪梨乾	2 兩	
南北杏	2 兩	
陳皮	1 個	# 浸軟
水	20 杯	
鹽	適量	# 後下

做法 @ 所有材料（除了鹽）置高身鍋內，先用大火煲滾 10 分鐘，改用中小火再煲 3 小時，飲用前先試味才下鹽。

comments...

此湯用蘋果吊出菜乾的香味 # 因菜乾屬瘦物，用略帶肥肉的不見天可滋潤湯水，飲時較香滑。

Pork and conch soup with dried Bok Choy and apples

雪梨乾在雜貨舖有售。

湯 soup

節瓜瑤柱螺頭湯

Gigi Wong 老火湯 ♥ 濃郁海味香 ✿ 好易煲 ♨

comments...

材料 @ 節瓜……1 斤 # 去皮，切環
瘦肉……半斤 # 汆水
瑤柱……4 粒 # 沖淨，略浸
乾螺頭……4 兩 # 處理好，處理方法看 P.89
紅棗……6 粒
水……18 杯
鹽……適量 # 後下

做法 @ 所有材料（除了鹽）放入鍋內，用猛火煲10分鐘後，改小火再煲3小時。
@ 飲用前先試味才下鹽。

comments...

Gigi Wong # 凡煮餸、煲湯，我都會先試味，將味道調校至最合心意才奉客。

湯 soup

★☆☆☆

土茯苓無花果蠔豉湯

Gigi Wong 有益健康 ♥ 按需要飲用 ✽ 容易烹調 ♨

宜選用美國出產、大粒的無花果。
這湯有收澀痔瘡、通便的功能。

這湯採用乾土茯苓，有除濕解毒之效；乾土茯苓在藥材舖有售。

Lean pork soup with Tu Fu Ling, dried figs and oysters

comments...

材料 @ 瘦肉……半斤 # 汆水
土茯苓……2 兩 # 浸透
蠔豉……6 兩 # 溫水浸透
淡菜……4 兩 # 溫水浸透
紫菜……2 兩 # 略浸去沙
無花果……10 粒
薑……4 片
水……18 杯
鹽……適量 # 後下

做法 @ 所有材料（除了鹽）放入鍋內，先用大火煲滾 10 分鐘，改用中慢火煲 3 小時。
@ 飲用前先試味才下鹽。

comments...

湯 soup

猴頭菇瘦肉湯

Gigi Wong 藥材老火湯 ♥ 容易入口 ✽ 趁熱飲用 ♨

這湯有預防胃酸倒流的功效。
猴頭菇味甘性平，有益脾胃、助消化之功效；白朮在湯內的作用是祛濕。

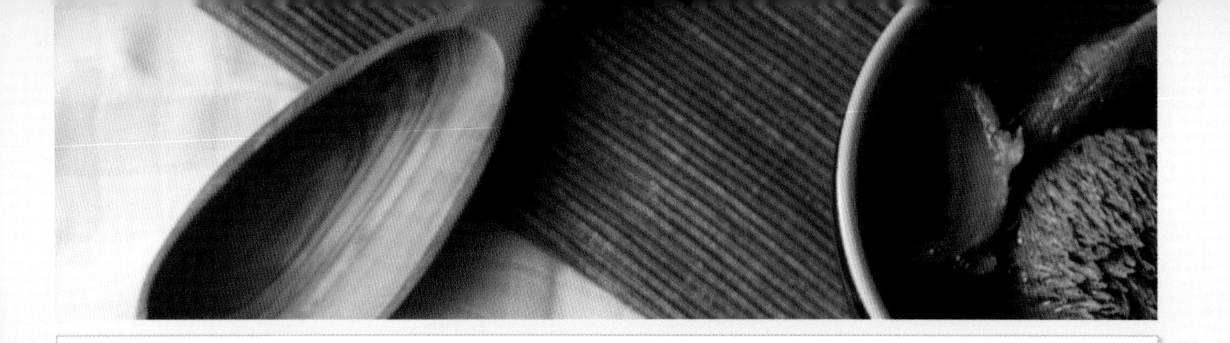

材料 @ 瘦肉............半斤 # 汆水
乾螺頭............2 兩 # 處理好，處理方法看 P.89
猴頭菇............2 兩 # 浸軟身
百合............2 兩 # 略浸
黨參............1 兩 # 沖洗
白朮............半兩
蜜棗............4 粒
陳皮............1 個 # 浸軟，刮去瓤
水............18 杯
鹽............適量 # 後下

做法 @ 所有材料（除了鹽）放入鍋內，先用大火煲約 10 分鐘，收中慢火煲 3 小時即可。
@ 飲用前先試味才下鹽。

comments...

湯 soup

老黃瓜養陰湯

Gigi Wong 老黃瓜老火湯 ♥ 祛濕養陰 ✽ 老幼皆宜 ♨

comments...

材料 @ 老黃瓜..........1 大條 # 洗淨，去瓤，連皮切大塊
牛蒡..........1 斤 # 去皮，切大塊
乾螺頭..........4 兩 # 處理好，處理方法看 P.89
雞腳..........4 對 # 汆水
花生、蓮子..........各 1 兩 # 沖洗，略浸
乾栗子..........1 兩 # 去殼，去衣
南北杏..........2 兩
紅棗..........8 粒 # 拍扁，去核
水..........20 杯
鹽..........適量 # 後下

做法 @ 所有材料（除了鹽）放入煲內，用大火煲滾 10 分鐘，再改中小火煲 3 小時。
@ 飲用前先試味才下鹽。

comments...

此湯在初夏飲用尤佳，有祛濕養陰的功效，加入栗子，有固腎之效。

紅棗：據說棗核很燥熱，我的習慣是將紅棗去核才煲湯。

湯 soup

★☆☆☆

土茯苓鑽地老鼠瘦肉湯

Gigi Wong 祛濕老火湯 ♥ 清熱解毒 ✽ 食療湯水 ♨

鑽地老鼠有清熱解毒、祛濕利水、消水腫的作用。如有體臭或腳臭的煩惱，皆因體內濕氣重，用鑽地老鼠煲湯飲用可幫助祛水、祛濕。
鑽地老鼠在街市有新鮮貨出售，也可在藥材舖購買乾貨。這湯用的是乾貨，如用鮮貨，則需要一個鑽地老鼠。

生薏米性涼，加入熟薏米有中和作用。至於洋薏米則有膠質，多用於煲糖水。

Lean pork soup with Tu Fu Ling and Mirabilis tuber

熟薏米

材料 @

瘦肉	半斤 # 汆水
乾螺頭	2 兩 # 處理好，處理方法看 P.89
土茯苓	2 兩 # 浸 2 小時後沖洗
乾鑽地老鼠	1 兩 # 沖洗
生、熟薏米	各 1 兩 # 沖洗
淮山	2 兩 # 沖洗
蓮子	半杯 # 沖洗
蜜棗	4 粒
陳皮	1 個 # 浸軟，刮去瓤
水	18 杯
鹽	適量 # 後下

做法 @ 所有材料（除了鹽）放入鍋內，先用大火煲滾 10 分鐘，改用中慢火煲 3 小時。飲用前先試味才下鹽。

comments...

湯 soup

牛蒡鯛魚湯

Gigi Wong 香甜鮮湯水 ♥ 有益健康 ✿ 老幼皆宜 ♨

這湯有寧心安神、減壓之效。

材料 @ 鯽魚 1 條 # 劏洗淨，煎至金黃色
牛蒡 1 條 # 去皮，切塊
粟米 2 條 # 切 4 節
紅蘿蔔 3 條 # 去皮，切塊
蓮子 1 兩 # 沖洗
腰果 1 兩 # 沖洗
杞子 1 兩 # 沖洗
百合 2 兩 # 沖洗
水 18 杯
鹽 適量 # 後下

做法 @ 鯽魚用煲湯魚袋裝好，與其餘材料（除了鹽外）放入鍋內，先用大火煲滾 10 分鐘，轉中慢火煲 3 小時。
@ 飲用前先試味才下鹽。

comments...

將鯽魚煎至金黃才煲湯，有助去除魚腥味。

Crucian carp soup with burdock

我煲魚湯，一定會用煲湯魚袋，不怕有鯁魚骨的危險，飲湯時安心又放心。

湯 soup

豬橫脷粉腸雞骨草湯

Gigi Wong 難度有 4 星 ♥ 雞骨草要洗淨 ✤ 適合要捱夜人士 ♨

這湯有清熱解毒、舒肝去瘀、去口苦，去煩熱的作用。
清洗雞骨草前，應先浸泡約 4 小時去掉泥沙；因為雞骨草附有很多泥沙和碎葉。

材料 @ 豬粉腸……半斤
豬橫脷……2 條 # 剪去脷底肥膏，汆水
雞骨草……半斤 # 沖洗乾淨
陳皮……1 個 # 浸軟，刮去瓤
蜜棗……8 粒
水……18 杯
鹽……適量 # 後下

做法 @ 豬粉腸套入去衣蒜頭，用手指將蒜頭擠壓由頭到尾貫通豬粉腸，做兩次，擠出腸內的污物。用生粉洗淨，過清水，汆水。
@ 所有材料（除了鹽）放入鍋內，先用大火煲滾 10 分鐘，改收中慢火再煲 3 小時。
@ 飲用前先試味才下鹽。

comments...

soup

螺頭石斛湯

Gigi Wong 滋陰安神 ♥ 明目 ✽ 老少咸宜 ♨

comments...

材料 @ 石斛……1 兩 # 略洗
乾螺頭……4 兩 # 處理好，處理方法看 P.89
瘦肉……半斤 # 汆水
淮山……2 兩 # 略洗
土茯苓……2 兩 # 略洗
黨參……1 兩 # 略洗
有皮蓮子……1 兩 # 略洗
杞子……1 兩 # 略洗
陳皮……1 個 # 浸軟，刮去瓤
水……20 杯
鹽……適量 # 後下

做法 @ 全部材料放入鍋內，先用大火煲滾 10 分鐘，改用中小火再煲 3 小時即可。
@ 飲用前先試味才下鹽。

comments...

Lean pork soup with dried conch and Shi Hu

湯 soup

★☆☆☆

節瓜章魚眉豆湯

Gigi Wong 湯味鮮香 ♥ 章魚是亮點 ✽ 好受歡迎 ♨

此湯能補氣生津、提神醒腦，對容易精神不振、疲倦乏力者有幫助。

勿去掉花生衣，花生連衣煲湯有補血的功效。

材料 @ 章魚乾......4 兩 # 浸軟
豬骨......半斤 # 汆水
大節瓜......2 個 # 去皮，切環
眉豆......1 杯 # 略浸
花生......1 杯 # 連衣略浸
薑......2 片
水......18 杯
鹽......適量 # 後下

做法 @ 所有材料（除了鹽）煲滾 10 分鐘後，改用中慢火再煲 3 小時即可。
@ 飲用前先試味才下鹽。

無論煲湯或作其他菜餚，通常都會撕去已浸軟的章魚膜，除了可去除澀味外，味道也比較濃郁。

節瓜要厚切才不易煲爛。

甜食 sweet

★☆☆☆

小鳳姐糖水

Gigi Wong 星級糖水 ♥ 好好味 ✽ 好易煮 ♨

這是小鳳姐（徐小鳳）的拿手甜品，材料只有寥寥四種，但味道有驚喜，神來之筆是最後下的牛油 # 用有鹽牛油會較香 # 在此與各位分享。

材料 @ 粟米粒……2 杯

牛奶……4 杯

糖……半杯

牛油……1 小塊

做法 @ 把粟米、牛奶及糖以中慢火煮至糖溶，邊煮邊攪，以防滾瀉。

@ 熄火前加入牛油即可享用。

comments...

甜食 sweet

薑片花生湯圓

Gigi Wong 甜蜜蜜 ♥ 薑味濃 ✲ 團圓必吃 ♨

將花生醬粒雪至硬，才容易用糯米皮包成湯圓。

湯圓材料 @ 糯米粉 1 杯
水 2/3 杯
花生醬 適量 # 搓成 12 小粒，放入冰格至硬

湯底材料 @ 薑 10 片
片糖 1 塊
水 2 杯

做法 @ 把水逐少注入糯米粉內，搓成粉糰，用濕毛巾蓋着半小時。
@ 湯底材料用中火煲至糖溶，薑出味。
@ 將糯米粉糰分成小粒，壓扁，包入花生醬，搓圓，放入滾湯內，煮至浮起即可。

comments...

餡料可自由配搭，選自己喜歡的，例如：豆沙、芝麻、片糖粒或椰絲砂糖等等。

Glutinous rice balls with peanut butter filling in ginger syrup

comments...

好立克曲奇

Gigi Wong horlicks 另一食法 ♥ 鬆脆 ✽ 烘出曲奇香 ♨

可以用三合一即溶咖啡或 Mocha 粉代替好立克。

材料 @ 自發粉............3/4 杯
好立克............3/4 杯 # 篩勻
無鹽牛油............150 克 # 室溫或「叮」20 秒
雞蛋............1 個 # 打散
雲呢拿香油............1 湯匙
杏仁碎............半杯

做法 @ 將全部材料（除了杏仁碎外）倒入大碗內，慢速打 2 分鐘。
@ 加入杏仁碎，用手搓勻成糰，用牛油紙捲起，置雪櫃 2 小時。
@ 取出粉糰並切成片，放在焗盤上，每塊之間要留些空位。
@ 預熱焗爐 170℃，先焗曲奇 10 分鐘，反轉曲奇再焗 8 分鐘即可，冷卻後才可放入瓶子內保存。

comments...

好立克粉容易受潮結成團，故定要用篩子篩去受潮粉末，烘後的曲奇味道才會均勻。

Horlicks cookies

comments...

Gigi Wong # 可以用芝麻代替杏仁碎。

comments...

甜食 sweet

芝麻花生糖

Gigi Wong 自製傳統糖果 ♥ 用優質麥芽糖 ✽ 香口好味 ♨

comments...

材料 @ 白芝麻............半斤 # 沖淨，瀝乾水分
紅衣花生............4 兩 # 沖淨，瀝乾水分
白糖............8 湯匙
麥芽糖............1 杯

做法 @ 先以中小火，用白鑊分別烘炒芝麻及花生，炒至花生微黃脆身，去衣。
@ 混合芝麻及花生，備用。
@ 易潔鍋內放白糖，小火加熱至微黃，開始溶化即可加入麥芽糖，快手攪拌直至濃稠，把芝麻花生料分兩次倒入，拌勻。
@ 趁熱倒入鋪了牛油紙的盤內，壓扁定型，冷卻後掰開享用。

comments...

如想芝麻花生糖外觀整齊，趁糖未硬化前切成自己喜愛的形狀。

Sesame peanut brittle

comments...

甜食 sweet

番薯玉米餅

Gigi Wong 健康粗糧 ♥ 微甜 ✽ 品嘗原味香 ♨

\# 這餅混合兩種粗糧而成，番薯不必搓成泥狀，以能夠食到番薯粗粒更佳。
\# 宜購買大小相近的番薯，因為容易控制「叮」熟的時間。

材料 A @ 番薯……675 克 # 洗淨，刺孔
牛油……110 克 # 室溫

材料 B @ 雞蛋……4 個 # 拌勻
牛奶……半杯

材料 C @ 自發粉……半杯 # 篩勻
玉米粉……1.5 杯
糖……1/4 杯
鹽……1 茶匙
肉桂粉……2 茶匙

做法 @ 番薯放入微波爐先「叮」5 分鐘，反轉再「叮」5 分鐘，取出去皮，趁熱加入牛油搓勻。
@ 材料 B 倒入番薯料內，用打蛋機拌勻。
@ 將材料 C 混合後，分兩次倒入以上的混合料內，充分拌勻即可倒入焗盤內。
@ 預熱焗爐 220℃，焗 45 分鐘即可。

comments...

Sweet potato cornbread

comments...

sweet

椰香奶凍

Gigi Wong 啖啖椰香 ♥ 口感軟滑 ✤ 透心涼 ♨

這奶凍口感軟滑，充滿椰香，是夏日的消暑佳品。

材料 @ 罐頭椰漿……100克
蘭姆酒（Rum）……20克
雞蛋……2個 # 拂勻
魚膠粉……1茶匙
糖……60克
忌廉……500克 # 打至企身

做法 @ 椰漿倒入大碗中，加入蘭姆酒、蛋液、魚膠粉及糖，最後加入已打起的忌廉，用打蛋器拌勻。
@ 將混合料倒入小碗中，放入雪櫃雪1小時，取出即可享用。

comments...

Sichuan "Ma La" cloud ear fungus_ 涼拌麻辣雲耳 P.8

INGREDIENTS:
1 cup cloud ear fungus

DRESSING:
1 tbsp "Ma La" seasoning
1 tbsp oyster sauce
1 tsp sugar
2 cloves garlic # finely chopped
1 sprig coriander # use stems only, finely chopped

METHOD:
1. Soak cloud ear fungus in water till soft. Cut off the hard stems. Blanch in boiling water for 5 minutes. Drain and let cool.
2. Mix the dressing ingredients together. Pour the cloud ear fungus into the dressing and toss to mix well. Serve.

GIGI'S NOTES:
Alternatively, you may put both the dressing and the cloud ear fungus into a zipper bag. Seal well and shake the bag to coat the cloud ear fungus evenly in the dressing. Pour into a serving bowl and serve.
In this recipe, I prefer using a smaller breed of cloud ear fungus, colloquially known as "rat ears." As they come in smaller florets, you don't need to tear them apart. That would save you some time and energy.
"Ma La" are the Chinese characters for numbing and spicy. The term refers to the tingling numbness in the mouth created by Sichuan peppercorns.

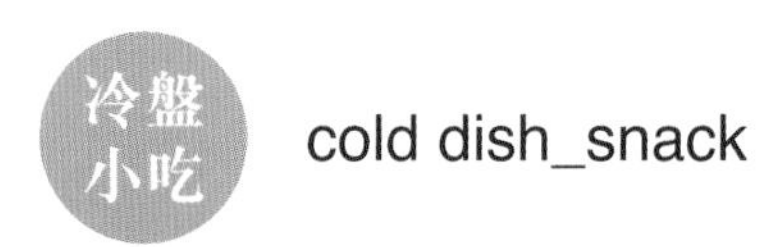

Warm potato salad in balsamic vinaigrette with fresh mint_ 薄荷薯仔 P.10

INGREDIENTS:
8 new potatoes

DRESSING:
2 sprigs fresh mint # leaves only, finely chopped
1 tsp salt
1 tbsp Balsamic vinegar
2 tbsp extra-virgin olive oil

METHOD:
1. Put new potatoes into a microwave-safe bowl. Add enough water to cover. Microwave over high power for 3 minutes.
2. Cut potatoes into chunks.
3. Mix all dressing ingredients. Put in the potatoes while still hot. Toss well and serve.

GIGI'S NOTES:
You may also put the potatoes and dressing into a zipper bag and shake it to coat the potatoes evenly in the dressing.

cold dish_snack

Beancurd stick cold appetizer_ 涼拌腐竹 P.12

INGREDIENTS:

3 beancurd sticks # soaked in water till soft, cut into short lengths, blanched in boiling water for 5 minutes
15 g wood ear fungus # soaked in water till soft, blanched in boiling water, finely shredded
50 g soybean sprouts # rinsed
2 slices ginger # finely shredded
1/2 red bell pepper # finely shredded

SEASONING:

1 tsp salt
1 tsp sesame oil

METHOD:

Heat a wok and add oil. Stir fry ginger and soybean sprouts over high heat. Add wood ear fungus, beancurd sticks and red bell pepper. Stir in seasoning and mix well. Transfer to plate. Let cool and serve.

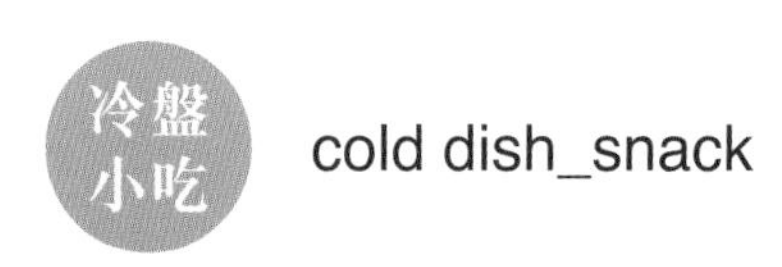

Flaky spring onion pancakes_ 香酥葱油餅 P.14

INGREDIENTS:
300 g round dumpling skin

FILLING:
1 cup diced spring onion
1 piece fatty pork # very finely chopped
1/2 tsp salt # added to fatty pork and chopped together

METHOD:
1. Spread a thin layer of chopped fatty pork over a piece of dumpling skin. Sprinkle with diced spring onion. Top with another piece of dumpling skin. Alternate with another layer of fatty pork and spring onion. Put another piece of dumpling skin on top.
2. Crimp the rim to fuse the edges of the dumpling skin.
3. Pour some oil in a pan. Fry the spring onion pancake over low heat until golden on both sides. Serve.

round dumpling skin

The fatty pork should be chopped as fine as this.

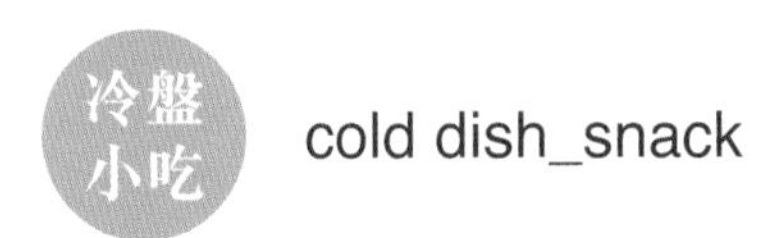

Sliced pork belly dressed in Sichuan garlic vinaigrette _ 蒜泥白肉 P.17

INGREDIENTS:
1 piece pork belly # about 225 g
1 cucumber # grated into long strips

AROMATICS:
2 thumbs ginger, 1 sprig spring onion, 2 tbsp Shaoxing wine

VINAIGRETTE DRESSING:
2 tbsp diced white part of spring onion # crushed with mortar and pestle
2 tbsp diced ginger # crushed with mortar and pestle
2 tbsp grated garlic # crushed with mortar and pestle
1 1/2 tbsp light soy sauce
1 tbsp Zhenjiang black vinegar
1 tbsp sugar
1 tsp sesame oil
1 tsp ground Sichuan peppercorns
1 tbsp chilli oil

METHOD:
1. Put pork belly and aromatics into a pot of cold water. Bring to the boil and cook for 15 minutes. Turn off the heat and cover the lid. Let sit for 5 minutes. Bring to the boil again and leave it to cool for 5 more minutes with the lid covered. Drain and soak the pork belly in ice water till completely cool. Refrigerate for 30 minutes.
2. Arrange the grated cucumber strips on a serving plate.
3. Slice the pork belly thinly and arrange over the bed of cucumber strips.
4. Drizzle with the mixed dressing all over. Serve.

GIGI'S NOTES:
I always soak the pork belly in ice water and refrigerate it for 30 minutes before slicing, so that the pork will be stiffer and easier to slice thinly.

rice_noodles

Rice vermicelli in herbal chicken soup and rice wine _ 雞酒米粉 P.20

INGREDIENTS:

1 bundle rice vermicelli # soaked in water briefly
2 boneless chicken thighs # cut into pieces, marinated
6 slices ginger
2 tbsp sesame oil
1/8 raw cane sugar slab
1/2 cup Hua Diao wine
1 tbsp Goji berries
1 slice Dang Gui
2 slices Huang Qi
2 tbsp grated garlic
2 cups boiling water

MARINADE FOR CHICKEN:

1 tbsp Shaoxing wine
1/2 tsp sugar
ground white pepper

METHOD:

1. Cook the rice vermicelli in boiling water until they loosen up. Drain and save in a large serving bowl.
2. In a pot, stir fry ginger in sesame oil until fragrant. Put in the chicken and stir to sear well. Add Hua Diao wine and bring to the boil. Put in the remaining ingredients. Cover the lid and cook for 5 minutes.
3. Optionally, add 2 tbsp of Hua Diao wine before serving. (Skip it if you are averse to the alcohol taste.) Pour the soup and chicken over the bowl of rice vermicelli. Serve.

Steamed millet_ 小米飯 P.22

INGREDIENTS:

1 cup millet
1 tbsp oil
Goji berries # rinsed

METHOD:

1. Soak millet overnight. Rinse and drain. Add oil to the millet and stir well to coat evenly.
2. Line a bamboo steamer with muslin cloth. Pour in the millet and steam over high heat for 30 minutes. Put in Goji berries before serving.

GIGI'S NOTES:

Millet helps the body recuperate, eases Asthenia, and strengthens vital energy flow in lower abdomen. It is suitable for those with physical weakness, Asthenia and poor appetite. Millet is harder than most grains. You must soak it in water overnight before steaming or making porridge with it. I prefer steaming it for its fluffy texture.

rice_noodles

Fried rice with dried radish_ 菜脯炒飯 P.24

INGREDIENTS:

2 eggs # whisked
4 bowls day-old rice
2 tbsp diced dried radish
2 Cantonese preserved pork sausages # diced
1/3 piece Cantonese preserved pork belly # diced
2 kale stems # diced
a few florets cloud ear fungus # soaked in water till soft; finely chopped
2 dried shiitake mushrooms # soaked in water till soft; diced
2 tbsp finely diced ginger
1 cup finely chopped spring onion

METHOD:

1. Heat a wok and add 1 tbsp of oil. Stir fry pork sausages and preserved pork belly over medium heat until oil is rendered. Add dried radish, kale, cloud ear fungus, shiitake mushrooms and diced ginger. Toss for 2 minutes. Set aside.
2. Add a little oil to the same wok. Pour in whisked eggs and stir until half set. Put in the rice. Stir to fluff up the rice and break the lumps. Put in the stir fried ingredients from step 1. Toss well and sprinkle with chopped spring onion. Toss again and serve.

GIGI'S NOTES:

It's wasteful to discard leftover rice especially when there is some left every day. I always keep my leftover rice in the freezer and make fried rice with it when it accumulates to a certain amount. It's yummy and it's a good way to reduce the impact of wasted food on the environment.

Stir fried beef with fresh cordyceps flowers
_ 鮮蟲草花炒牛肉 P.26

INGREDIENTS:
225 g beef # thinly sliced, marinated
120 g fresh cordyceps flowers # rinsed, roots trimmed off
1/2 onion # shredded

AROMATICS:
2 shallots # finely chopped
4 slices ginger
2 sprigs spring onion # use white part only, cut into short lengths

MARINADE FOR BEEF:
1/4 tsp sugar
1/2 tsp salt
1/2 tsp ground white pepper
1 tsp caltrop starch slurry # 1 tsp caltrop starch mixed well with 1 tbsp water
1 tsp oil

SEASONING:
1 tsp salt
1 tbsp oyster sauce

METHOD:
1. Heat 1 tbsp of oil in a wok. Stir fry aromatics until fragrant. Put in the marinated beef and toss quickly for 10 seconds. Set aside.
2. In the same wok, heat some oil. Stir fry onion and cordyceps flowers over high heat. Put in the beef. Add seasoning and toss well. Serve.

GIGI'S NOTES:
You may use dried or fresh cordyceps flowers for this recipe. If you use dried ones, soak them in water until soft and drain before using. Fresh cordyceps flowers are crisper in texture whereas dried ones are chewier.

Chinese surf-and-turf stew_ 全家福 P.29

INGREDIENTS:

1 chicken # dressed and chopped into pieces
1 large slab pork ribs # chopped into chunks
1 pork tripe # rinsed, cut into pieces
1 cuttlefish # removed membrane, cut into pieces
(Put together in a bowl and tossed with marinade)
4 eggs, flour, wine
8 red dates # stoned

MARINADE:

1 tsp salt, 1/2 tsp sugar, 1 tbsp caltrop starch
1 tsp ground white pepper, 1 tsp sesame oil

MEATBALL INGREDIENTS:

2 dried shiitake mushrooms # soaked in water till soft, drained, finely chopped
300 g ground pork # marinated
1 whisked egg
plain flour

AROMATICS:

4 cloves garlic, 2 slices ginger
1 bird's eye chilli, 2 cubes rock sugar
1 whole pod star anise

SEASONING:

1 tbsp light soy sauce
1 tbsp dark soy sauce
1 tbsp oyster sauce
1 tsp salt

METHOD:

1. In a mixing bowl, put in chopped shiitake mushrooms and ground pork. Mix well and shape into meatballs. Dip each meatball into whisked egg and coat it in flour. Deep-fry in oil until golden. Drain and set aside.
2. Put the eggs into a pot of cold water. Bring to the boil and cook for 5 minutes until hard boiled. Shell them and coat them lightly in flour. Deep-fry in oil briefly until golden. Set aside.
3. Heat a wok. Pour in cold oil. Stir fry aromatics until fragrant. Add chicken, pork ribs, pork tripe and cuttlefish. Sizzle with wine. Add seasoning and toss well. Add water to cover all ingredients. Put in red dates. Bring to the boil and turn to low heat. Simmer for 30 minutes.
4. Put in the meatballs and eggs. Cook until the sauce reduces. Serve.

GIGI'S NOTES:

1. You may get readymade pork balls from the market. Then you can save time and energy on making them from scratch.
2. I coated the hard boiled eggs in flour lightly and deep-fried them. Not only do they look golden and gorgeous this way, they also tend to hold their shapes better after prolonged cooking.

小菜 main dish

Pan-seared spareribs_ 生炒金沙骨 P.32

INGREDIENTS:

pork spareribs # cut into pieces

SEASONING:

2 cubes rock sugar, 2 tbsp fish sauce
1/2 cup water # added little by little depending on how quickly the sauce dries out

METHOD:

1. Sear the spareribs in little oil until they turn white on all sides.
2. Add rock sugar and fry until the ribs are lightly browned. Add fish sauce and water. Bring to the boil. Cover the lid and cook over low heat until the sauce reduces. Serve.

main dish

Steamed eggplant with sesame paste and garlic _ 麻醬蒜泥蒸茄子 P.86

INGREDIENTS:

2 eggplants # cut into short lengths, with light crisscross incisions made on the skin
flour

DRESSING:

2 tbsp sesame paste
2 tbsp Zhenjiang black vinegar
1 clove garlic # grated
1/2 tsp salt, 1 tsp sugar

METHOD:

1. Sprinkle flour lightly on the eggplant. Save on a plate. Steam for 7 minutes.
2. Dribble the dressing evenly on the eggplant. Serve.

Steamed pork trotter in red yeast rice paste
_ 紅麴豬手 P.34

INGREDIENTS:
1 pork trotter # chopped into pieces; marinated
2 tbsp water
2 tbsp red yeast rice
1 tbsp glutinous rice flour
1/2 cup peanuts # with pink skin on; soaked in water for 30 minutes

MARINADE FOR PORK TROTTER:
1 tsp salt
1 tbsp light soy sauce
ground white pepper
2 tbsp Shaoxing wine
2 tsp brown sugar

METHOD:
1. Put pork trotter into a mixing bowl. Add remaining ingredients and mix to coat evenly. Transfer onto a steaming plate. Steam for 45 minutes.
2. Flip each piece of pork trotter over and steam for 45 more minutes. Serve.

GIGI'S NOTES:
I steam the pork trotter for a total of 90 minutes for tender and soft texture. If you prefer some crunch in your pork trotter, just steam it for 60 minutes.
I steam the pork trotter for 45 minutes on each side so that each piece will be coloured more evenly.

main dish

Homestyle stir fried ground pork and diced kale
_ 鄉里炒粒粒 P.36

INGREDIENTS:
200 g ground pork # marinated
2 tbsp pork cracklings # chopped
1 cup diced kale stems
2 tbsp fermented black beans # soaked in warm water; drained
1 red chilli # diced
1 thousand-year egg # shelled and diced
2 cloves garlic
2 slices ginger

MARINADE FOR GROUND PORK:
1/2 tsp salt
ground white pepper

SEASONING:
1 tsp sugar
1 tbsp oyster sauce

METHOD:
1. Stir fry garlic in a little oil until fragrant. Stir fry ground pork until half-cooked. Set aside.
2. In the same wok, stir fry ginger in the remaining oil. Put in diced kale stems and toss well. Add fermented black beans, red chilli and the half-cooked ground pork from step 1. Toss again. Put in pork cracklings and diced thousand-year egg at last. Add seasoning and stir to mix well. Serve.

GIGI'S NOTES:
Instead of pork cracklings, you may use fatty pork for similar effect. Just dice it and deep fry until lightly browned.
The original recipe calls for diced radish greens. In the old days, folks from the countryside wouldn't let any edible ingredient like radish greens go wasted. Nowadays, the white radish we see in the market comes with the greens removed. That's why I settle for kale stems instead.

Chitterlings stew with salted mustard greens
_ 鹹酸菜焖豬大腸 P.39

INGREDIENTS:
2 sets frozen pork large intestines # about 1.2 kg; blanched in boiling water for 1 hour
1 head salted mustard greens # sliced

BLANCHING STOCK FOR CHITTERLINGS:
1 tbsp white peppercorns
2 slices ginger
1 leek
2 tbsp rice wine

AROMATICS:
6 slices ginger
4 sprigs spring onion # cut into short lengths
2 sprigs coriander # cut into short lengths
1 dried red chilli # sliced diagonally

SPICES:
2 amomum tsao-ko
2 slices liquorice
2 bay leaves
2 whole pods star anise
1/3 dried tangerine peel
2 tsp Sichuan peppercorns # put into muslin bag; tied well

SEASONING:
2 tbsp chilli bean sauce
2 tbsp oyster sauce
1 tbsp light soy sauce
1 tbsp white vinegar
1 tsp sugar
1 tsp ground white pepper
1 cup boiling water

METHOD:
1. Tear off a small piece of mustard greens and taste it. If it's too salty, soak it in salted water for 30 minutes to partly draw out the salt. Slice the mustard greens. Fry it in a dry wok until it dries up. Turn to low heat and add 2 tsp of sugar and 1 tsp of oil. Toss well and set aside.
2. Trim off any fat on the chitterlings. Put them into a pot and add blanching stock ingredients. Add enough water to cover. Bring to the boil and turn to medium heat. Cook for 1 hour. Drain. Slice the chitterlings.
3. Heat a wok and add cold oil. Stir-fry aromatics in the order listed until fragrant. Turn to low heat. Put in the spices. Stir until fragrant. Discard the bag of Sichuan peppercorns. Put in chitterlings and sliced salted mustard greens. Add seasoning and cook over low heat for 10 minutes. Serve.

GIGI'S NOTES:
In step 1, I turned the heat down to low before adding sugar because sugar burns very easily. Burnt sugar would make the mustard greens taste bitter. Adding oil help greasing the mustard greens. It tastes richer after frying in a dry wok and tossed in sugar and oil.

main dish

Braised pork trotter_ 燉元蹄 P.42

INGREDIENTS:

1 pork trotter about 1.2 kg # blanched in boiling water
600 g leafy greens # rinsed
4 slices ginger
4 sprigs spring onion # each tied into a knot
8 red dates # stoned

SEASONING:

2 tbsp Shaoxing wine
1 tbsp dark soy sauce
2 tbsp light soy sauce
3 cubes rock sugar # crushed
1 tbsp red yeast rice # wrapped in muslin or teabag

BLANCHING STOCK FOR LEAFY GREENS:

1 tbsp oil
1 tsp salt
2 cups water

METHOD:

1. Dry the blanched pork trotter thoroughly with paper towel. Heat a wok and add oil. Fry the pork trotter on all sides until golden. Set aside.
2. Drain oil from the wok, saving a little for frying ginger and spring onion. Put in ginger and spring onion and fry until fragrant. Put in seasoning in the exact order listed. Add red dates and pour enough water to cover. Put in the pork trotter. Keep rolling it in the mixture to colour it evenly.
3. Transfer the pork trotter with all aromatics and sauces into a pressure cooker. Heat for 30 minutes. Check if there is much sauce left. Heat over high heat to reduce the sauce if necessary. Keep drizzling the sauce over the pork trotter to colour it and season it.
4. Boil the blanching stock for leafy greens. Put in leafy greens and cover the lid. Cook for 2 minutes. Drain and arrange on a serving plate. Put the braised pork trotter over the bed of leafy greens. Dribble the sauce over the pork trotter. Serve.

Honey-glazed barbecue pork cheek_ 蜜汁豬頸肉 P.44

INGREDIENTS:
4 pork cheeks # marinated overnight

MARINADE:
210 ml teriyaki sauce
1/2 cup Shaoxing wine
1 tsp salt

HONEY GLAZE:
3 tbsp honey
2 tbsp black pepper sauce
1 tbsp Worcestershire sauce

METHOD:
1. Heat some oil in a pan. Sear both sides of each pork cheek over low heat until lightly browned and cooked through. Pour in honey glaze. Cook until the glaze reduces. Serve.
2. Alternatively, you can grill the pork cheeks in an oven instead. Preheat an oven with only upper heat filaments on up to 230°C. Grill each side of the pork cheek for 10 minutes. Brush on honey glaze. Turn oven down to 180°C. Grill each side for 2 minutes. Serve.

GIGI'S NOTES:
Honey burns easily when subject to high heat. Thus, make sure the meat is cooked through before adding honey. Cook honey over medium heat at most.
Teriyaki sauce is a Japanese condiment. You may also use Chu Hau Sauce for similar results.

Stir fried pork belly with yam bean_ 沙葛炒五花腩 P.47

INGREDIENTS:

1/2 yam bean about 350 g # peeled; shredded; blanched in boiling water
1/2 salted mustard greens # shredded; soaked in water for 15 minutes; drained
200 g pork belly # skinned; sliced; marinated
90 g flowering Chinese chives # cut into short lengths
1 bird's eye chilli # shredded
2 slices ginger # shredded
2 cloves garlic # diced
1 tbsp Shaoxing wine

MARINADE FOR PORK:

1/2 tsp salt
ground white pepper

SEASONING:

1/2 tsp salt
1/2 tsp sugar
1 tsp caltrop starch
1 tbsp oyster sauce
1 tbsp water
ground white pepper

METHOD:

1. Fry the salted mustard greens in a dry wok over high heat for about 5 minutes. Turn to low heat and add 2 tsp of sugar. Toss well. Add 1 tsp of oil to grease it. Set aside.
2. Heat a wok and add oil. Stir-fry ginger and garlic till fragrant. Toss in pork belly slices and flip them to separate. Sizzle with wine. Put in yam bean and salted mustard greens. Stir briefly and add seasoning. Put in flowering Chinese chives and bird's eye chilli at last. Toss a few times. Serve.

GIGI'S NOTES:

To slice the pork belly neatly and easily, you may keep it in the freezer briefly to stiffen it.
Here's a trick on how to peel yam bean. Just make a light incision on the top tip. Then tear the skin off from the cut.

main dish

Steamed pork ribs with pickled chopped chillies
_ 剁椒蒸排骨 P.50

INGREDIENTS:
450 g pork spareribs # cut into chunks
1/2 taro # peeled and sliced
1/2 tomato # diced finely

MARINADE FOR PORK RIBS:
1 tsp salt
2 tsp sugar
1 tsp sesame oil
1 tsp dark soy sauce
1 tbsp oyster sauce
1 tbsp fermented black beans
1 tbsp pickled chopped chillies
2 tsp cornstarch

METHOD:
1. Add 2 tsp of salt to the pork ribs. Rub to coat them in salt evenly. Rinse well. Add 2 tsp of salt and rub again. Rinse. Wipe dry with paper towel. Add marinade and mix well.
2. Line the bottom of the dish with sliced taro. Arrange the marinated pork ribs on top.
3. Top with diced tomato. Steam over high heat for 12 minutes. Serve.

GIGI'S NOTES:
The diced tomato not only adds a lovely colour to the dish, but also partly neutralizes the piquancy of the pickled chillies.

Braised beef shin in aromatic soybean paste
_ 紅煨牛肉 P.52

INGREDIENTS:
500 g beef shin
2 sprigs Peking scallions # sliced diagonally into short lengths

AROMATICS:
2 cubes rock sugar
2 slices ginger
2 cloves garlic
1 whole pod star anise
1 red chilli

SEASONING:
1 tbsp ground soybean paste
2 tbsp Shaoxing wine
1 tbsp dark soy sauce
1/2 tsp salt

METHOD:
1. In a pot of cold water, put in 2 slices of ginger and the beef shin. Bring to the boil and skim off the foamy scum on top. Blanch the beef for 30 minutes. Strain the stock for later use.
2. Slice the beef shin thickly.
3. Heat some oil in another pot over low heat. Stir fry aromatics until fragrant. Put in the beef. Pour in the beef stock from step 1 to cover the beef shin. Bring to the boil and turn to medium-low heat. Simmer for 1 hour until the beef shin is tender.
4. Fry Peking scallions in a little oil until lightly browned. Arrange along the rim of a serving plate. Put the beef shin at the centre. Dribble with some sauce. Serve.

GIGI'S NOTES:
When you blanch the beef shin, putting ginger in the cold water helps to remove the gamey taste. The foamy scum on the surface must be skimmed off. This beef stock should be used to cook the beef again. Cooking meat in its own juices is a new skill I acquired recently.
The red chilli adds another dimension to this dish. You can taste the mild piquancy in the background and it's very appetizing.

After hours of cooking, the onion is caramelized and picks up the meaty flavour. It tastes extremely good.

In the photo, it may seem a little violent to grab the duck's neck and roll it around in the hot oil to sear the skin golden. But I find it the easiest way to handle it and it works best. Give it a try.

main dish

Braised duck in onion sauce_ 洋葱鴨 P.55

INGREDIENTS:
1 duck # rinsed; wiped dry
8 onions # cut into wedges

SEASONING:
5 tbsp dark soy sauce
4 tbsp light soy sauce
3 tbsp Shaoxing wine
2 tbsp sugar
1 cup water

METHOD:
1. Rub 3 tbsp of dark soy sauce on the skin of the duck evenly. Let dry. Fry the duck in oil until golden on all sides. Set aside.
2. Stuff the duck with onion wedges. Then line the bottom of a heavy pot with the remaining onion wedges. Put the duck over the bed of onion. Drizzle with seasoning and cover the lid. Cook over medium-low heat for 60 to 90 minutes until the duck is tender. Serve.

GIGI'S NOTES:
The seasoning is easy to remember as it's ordered from 1 to 5.
After hours of cooking, the onion is caramelized and picks up the meaty flavour. It tastes extremely good.

main dish

Duck stew with winter melon_ 冬瓜燜鴨 P.58

INGREDIENTS:
1 dressed duck # chopped into pieces, marinated
600 g winter melon # de-seeded, cut into chunks
1 thumb ginger # lightly crushed
4 cloves garlic
75 g fox nuts # rinsed, soaked in water briefly
75 g Job's tears # rinsed, soaked in water briefly
1 piece dried tangerine peel # soaked in water till soft, with the pith scraped off

MARINADE FOR DUCK:
1 tsp salt
ground white pepper

SEASONING:
1 tbsp oyster sauce
1 tsp salt

METHOD:
1. Heat a wok till very hot. Add cold oil. Stir fry ginger and garlic until fragrant. Put in the duck pieces. Add fox nuts, Job's tears and dried tangerine peel. Add water to cover. Bring to the boil over high heat. Turn to medium heat and stew for 45 minutes.
2. Put in the winter melon. Cook for 30 more minutes. Add seasoning. Mix well and serve.

GIGI'S NOTES:
In this recipe, do not season with light soy sauce. Otherwise, it would taste sour.
Ask the butcher to pick a young duck for this dish. The flesh of young ducks does not taste as flavourful as that of mature duck, but it is tenderer and less rubbery. Young ducks work well in stews or braised recipes.

Stir fried chicken strips with cucumber
_ 青瓜炒雞柳 P.60

INGREDIENTS:
2 cucumbers # sliced thickly
1 boneless chicken thigh # cut into thick strips, marinated
1 red bell pepper # cut into thick strips
1 sprig spring onion # use white part only, cut into short lengths
2 cloves garlic

MARINADE:
1 tsp salt
1 tsp ground black pepper

SEASONING:
1 tbsp oyster sauce
1/2 tsp sugar
1 tsp sesame oil

METHOD:
1. Put the sliced cucumbers into a plastic bag. Add 1 tsp of salt. Seal tightly. Then shake forcefully to draw moisture out of the cucumbers. Rinse with cold water and squeeze dry.
2. Heat a wok and add 1 tbsp of oil. Stir fry garlic until fragrant. Then stir fry chicken until it turns white. Remove chicken. In the same wok, put in cucumber and stir fry. Add chicken and toss well. Then put in red bell pepper and white part of spring onion. Toss briefly. Add seasoning and stir well. Serve.

GIGI'S NOTES:
There are different varieties of cucumbers in the market and for this recipe, I prefer those with bumps on the skin to the smooth ones. The cucumber tastes crunchier after shaken with salt without the extra moisture that thins out the sauce.
Instead of chicken thigh, you may use pork cheek, shelled shrimps or even dried shrimps for this recipe. They work differently, yet equally tasty.
Those who prefer more fiery heat in their food may use bird's eye chillies instead of red bell pepper.

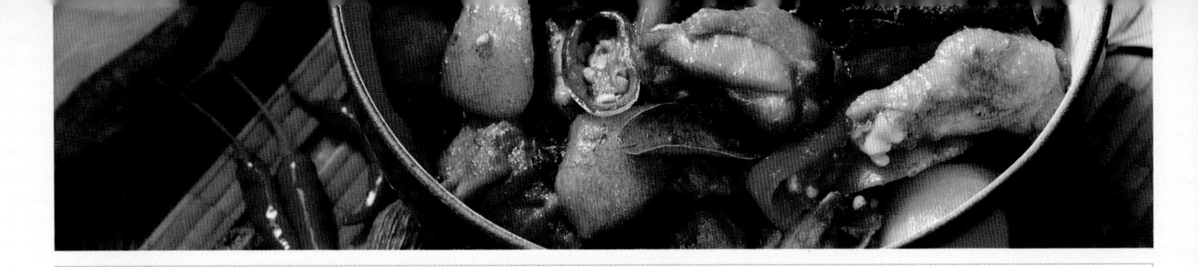

main dish

Beer braised chicken with Sichuan peppercorns and spices_ 大盤雞 P.63

INGREDIENTS:
1 chicken # chopped into pieces, marinated
1 tbsp Sichuan peppercorns
4 potatoes # peeled and cut into chunks
1 small bottle beer
1 green bell pepper # cut into chunks
1 red bell pepper # cut into chunks

MARINADE FOR CHICKEN:
1 tsp salt
ground white pepper

AROMATICS:
2 cubes rock sugar
4 slices ginger
4 cloves garlic
4 sprigs spring onion # cut into short lengths
1 amomum tsao-ko
1 whole pod star anise
1 piece cassia bark
2 bay leaves

SEASONING:
2 tbsp dark soy sauce
1 tbsp light soy sauce
2 tsp salt

METHOD:
1. Heat oil in a wok. Stir fry Sichuan peppercorns until fragrant. Remove and discard.
2. In the same wok, stir fry aromatics until fragrant in the Sichuan pepper oil. Put in the chicken pieces. Toss until the chicken turns white. Add potatoes and beer. Cook for 15 minutes or until the sauce reduces.
3. Add bell peppers. Toss well. Add seasoning and toss again. Serve.

 main dish

Stir fried soybean sprouts with salmon
_ 三文魚炒大豆芽 P.66

INGREDIENTS:

1 salmon fillet about 200 g # cut into thick strips with skin on, marinated
4 slices ginger # shredded
160 g soybean sprouts, 1 pack bonito flakes
4 fresh shiitake mushrooms # shredded
20 g wood ear fungus # soaked in water till soft, shredded

MARINADE FOR SALMON:

1/2 tsp salt
ground white pepper

SEASONING:

1/2 tsp salt
1 tbsp rice wine

METHOD:

1. Heat a wok and add some oil. Fry the salmon strips over medium heat until oil is rendered. Set aside.
2. Stir fry shredded ginger with the remaining oil in the wok until fragrant. Add soybean sprouts, shiitake mushrooms and wood ear fungus. Toss well. Sizzle with wine. Season with salt. Put the salmon strips from step 1 back in the wok. Toss well. Sprinkle with bonito flakes. Serve.

 # Salmon skin is rich in fish oil. You just need a little oil when you stir fry the salmon strips.

comments...

小菜 main dish

Fried prawns in Sichuan Yu Xiang sauce_ 魚香蝦球 P.68

INGREDIENTS:
10 medium prawns
caltrop starch
1 tbsp broad bean paste
2 tbsp Shaoxing wine

AROMATICS:
2 bird's eye chillies # finely chopped
2 sprigs spring onion # white parts only, finely chopped
4 slices ginger # finely diced
4 cloves garlic # finely diced

SICHUAN YU XIANG SAUCE:
1 tbsp caltrop starch
2 tbsp light soy sauce
3 tbsp sugar
4 tbsp Shanxi aged vinegar

METHOD:

1. Shell the prawns. Rub salt on them and rinse. Repeat rubbing salt and rinsing once more. Wipe dry with paper towel. Make a cut only the spine of each prawn. Devein.
2. Coat the prawns lightly in caltrop starch. Fry in a little oil until medium-well done. Set aside.
3. In a wok, heat some oil and fry aromatics until fragrant. Add broad bean paste and stir well. Sizzle with wine. Turn to high heat and pour in the Sichuan Yu Xian sauce ingredients. Bring to the boil. Put in the prawns. Toss well. Serve.

main dish

Steamed grey mullet in salted lemon paste

_ 鹹檸檬醬蒸魚 P.71

INGREDIENTS:

1 grey mullet about 900 g # dressed with gills removed; scaled; belly cut open
2 sprigs spring onion # cut into halves
1 bird's eye chilli # finely shredded

SALTED LEMON PASTE:

4 slices pickled ginger # finely shredded
1 salted lemon # finely shredded
3 pickled plums # mashed; stoned
4 tbsp sour plum sauce
8 pickled Chinese bulbous onions # finely shredded
1 tbsp sugar
1/2 cup water

METHOD:

1. Put spring onion evenly on a steaming plate. Put the fish on the bed of spring onion by lifting both sides of the cut belly. It should stand upright on its belly.
2. Steam over high heat for 6 minutes until done. Remove the spring onion.
3. Cook the salted lemon paste until flavours blend in. Add shredded bird's eye chilli. Drizzle over the steamed fish. Serve.

main dish

King oyster mushrooms in cinnamon glaze
_ 肉桂雞髀菇 P.74

INGREDIENTS:
500 g king oyster mushrooms # wiped down, roll cut into pieces
4 cloves garlic # finely chopped
1 red chilli # finely chopped
2 sprigs spring onion # finely chopped
1 sprig coriander # finely chopped

SEASONING:
1 cup water
2 tbsp brown sugar
1 tbsp white sugar
1 tsp ground cinnamon
1 tbsp dark soy sauce

METHOD:
1. Fry the mushrooms in a dry wok for 3 minutes. This helps dry the mushrooms.
2. Heat a wok and add 1 tbsp of oil. Stir fry garlic and red chilli until fragrant. Pour in seasoning in the order listed. Cook until the glaze thickens. Pour in the mushrooms. Toss well. Add spring onion and coriander. Toss again. Serve.

main dish

Three-cup assorted mushrooms_ 三杯菇 P.76

INGREDIENTS:
4 king oyster mushrooms # sliced
4 fresh shiitake mushrooms # cut in halves
1 pack Hon-shimeji mushrooms
1 tbsp sesame oil
1 tsp vegetable oil
1 cup Thai basil leaves # added last

AROMATICS:
1/2 onion # shredded
2 slices ginger
2 cloves garlic
1 red chilli

SEASONING:
1 tbsp black vinegar
1 tsp brown sugar

METHOD:
1. Stir fry aromatics in vegetable oil and sesame oil. Put in all three kinds of mushrooms. Toss for a few minutes to sweat them.
2. Add seasoning and stir well. Put in Thai basil leaves. Quickly toss to mix well. Serve.

GIGI'S NOTES:
You may use any mushroom in season for this recipe. Before using, just wipe them down with damp cloth. Do not rinse them in water. Otherwise, they will suck up all water like a sponge and the dish would end up watery.
Sesame oil tends to burn more easily than vegetable oil. Thus, in this recipe, I use both to fry the mushrooms which will pick up the aroma of sesame oil without burning.

main dish

Stir fried fresh prawns with Chinese chives
_ 韭菜鮮蝦 P.78

INGREDIENTS:

200 g fresh prawns # antennae and legs trimmed off
100 g Chinese chives # rinsed and cut into short lengths
2 bird's eye chillies # finely chopped
30 g perilla leaves # rinsed and finely chopped

SEASONING:

1/2 tsp sugar, 1 tsp salt
1 tsp sesame oil, ground white pepper

METHOD:

1. Blanch the prawns in oil (there should be more oil than shallow frying, but less oil than deep frying). Stir until the prawns turn red. Set aside and drain most of the oil.
2. In the same wok, fry the bird's eye chillies over high heat until fragrant. Put in the prawns, Chinese chives and perilla leaves. Add seasoning and toss to coat evenly. Serve.

Napa cabbage rolls with minced shrimp filling
_ 黃芽白包釀蝦膠 P.80

INGREDIENTS:
300 g shelled shrimps
75 g fatty pork # finely chopped
12 Napa cabbage leaves # blanched in boiling water for 2 minutes
1 1/2 egg whites # whisked

SEASONING FOR MINCED SHRIMPS:
1/2 tsp ground white pepper
1/2 egg white
1/2 tsp sesame oil
1 tsp Shaoxing wine

GLAZE:
1 tsp caltrop starch
1 tbsp water
1 tsp sesame oil
1/2 tsp salt
1/2 cup juices from steaming Napa cabbage rolls

METHOD:
1. Rub salt on the shelled shrimps evenly. Rinse. Repeat rubbing salt and rinsing one more time. Wipe dry with paper towel. Crush with the flat side of a knife and chop to mince well.
2. Add seasoning to the minced shrimps. Mix well and leave it for 30 minutes. Put in fatty pork. Stir in one direction until sticky. Slap the minced shrimp mixture onto a chopping board a few times.
3. Wipe dry the blanched Napa cabbage leaves. Lay one flat on a chopping board. Spread some minced shrimp over it and fold into a packet. Arrange on a steaming plate.
4. Steam over high heat for 5 minutes. Drain any liquid on the plate and use it to make the glaze. Stir well the glaze ingredients and heat it up until it thickens. Turn off the heat and stir in the egg white. Pour this glaze over the steamed Napa cabbage rolls. Serve.

GIGI'S NOTE:
You may use white cabbage instead of Napa cabbage, but I found Napa cabbage tastes sweeter.
Stir minced shrimp in one direction. It turns sticky more quickly that way.

main dish

Sizzling grass carp fillet in clay pot_ 啫啫鯇魚塊 P.83

INGREDIENTS:
1 big piece grass carp belly
1 tbsp Shaoxing wine

MARINADE FOR FISH:
1 tsp salt
ground white pepper
1 tsp caltrop starch
4 preserved black olives # chopped, with 1 tsp of sugar added and steamed for 10 minutes
1 tsp sesame oil

AROMATICS:
8 slices ginger, 8 slices galangal
8 cloves garlic, 8 shallots
2 sprigs spring onion # white part only, cut into short lengths

METHOD:
1. Cut the grass carp belly into thick strips. Add marinade and mix well. Leave them for 1 hour.
2. Heat 2 tbsp of oil in a clay pot. Stir fry the aromatics until fragrant. Arrange the marinated fish fillet over the aromatics.
3. Cover the lid and slowly pour Shaoxing wine along the rim of the lid. Cook for 6 minutes. Serve the whole pot.

GIGI'S NOTES:
Pouring the Shaoxing wine along the rim of the lid lets the fish pick up the aroma evenly.

湯 soup

Nourishing herbal soup for autumn_ 秋天滋潤湯 P.88

INGREDIENTS:

38 g dried pipefish # rinsed
38 g Sha Shen # rinsed
38 g Yu Zhu # rinsed
38 g dried lily bulb # rinsed
38 g almonds # rinsed
38 g Huai Shan # soaked in water and rinsed
38 g black beans # soaked in water for 30 minutes, drained
150 g dried conch # see Gigi's notes for preparation
38 g dried figs # rinsed
1 whole dried tangerine peel # soaked in water till soft
1 carrot # peeled and cut into chunks
18 cups water
salt # to taste; added last

METHOD:

1. Put all ingredients (except salt) into a pot. Bring to the boil over high heat. Cook for 10 minutes. Turn to medium-low heat. Simmer for 3 hours.
2. Taste the soup before seasoning with salt.

GIGI'S NOTES:

To prepare the dried conch, rinse 600 g of dried conch. Add enough water to cover the conch and soak for overnight. Pour the conch and water into a rice cooker. Put in a cube of rock sugar. Turn on the rice cooker and cook until water dries up. Let cool and divide the conch among several zipper bags. Keep in the freezer and use it whenever a recipe calls for dried conch. It saves time and much effort this way.

soup

Fish soup with peanuts and papaya_ 花生木瓜魚仔湯 P.90

INGREDIENTS:

600 g any small fish # dressed, fried in oil till golden, put into a muslin bag and tied tightly
1 cup peanuts # rinsed, with skin on
1 mature papaya # peeled, de-seeded, cut into pieces
1/3 cup cashew nuts # rinsed
1/3 cup chestnuts # rinsed
15 red dates # crushed, stoned
1 whole dried tangerine peel # soaked in water till soft, with pith scraped off
18 cups water
salt # to taste; added last

METHOD:

1. Put all ingredients (except salt) into a pot. Boil over high heat for 10 minutes. Turn to medium-low heat. Simmer for 3 hours.
2. Taste the soup. Season with salt accordingly.

GIGI'S NOTES:

This soup is inexpensive and easy to make. It strengthens the body and is great for all ages.
The pink skin of peanuts is effective in regenerating blood cells. Thus, do not skin the peanuts. A tea made by boiling peanut skin in water is believed to help blood cell regeneration among cancer patients after chemotherapy.
You may use a fish tail instead of small fish.

湯 soup

Pork and conch soup with dried Bok Choy and apples _ 菜乾蘋果湯 P.94

INGREDIENTS:

600 g pork inner shoulder butt # blanched in boiling water, drained
150 g dried conch # see p.159 for preparation
150 g dried Bok Choy # soaked in water until soft, rinsed, with sand removed
4 apples # halved, cored
75 g dried Ya-li pear
75 g sweet and bitter almonds
1 whole dried tangerine peel # soaked in water till soft
20 cups water
salt # to taste; added last

METHOD:

Put all ingredients (except salt) into a tall stock pot. Boil over high heat for 10 minutes. Turn to medium-low heat and simmer for 3 hours. Taste the soup and season with salt accordingly.

GIGI'S NOTES:

The apples heighten the flavour of dried Bok Choy in this soup. As dried Bok Choy tends to pick up any grease in the soup, using the inner shoulder butt cut of pork adds more oil to the soup, so that it tastes richer.

湯 soup

Lean pork soup with Chinese marrows, dried scallops and conch_ 節瓜瑤柱螺頭湯 P.96

INGREDIENTS:

600 g Chinese marrows # peeled; cut into chunks
300 g lean pork # blanched in boiling water
4 dried scallops # rinsed and soaked in water briefly
150 g dried conch # see p.159 for preparation
6 red dates, 18 cups water
salt # to taste; added last

METHOD:

1. Put all ingredients (except salt) into a pot. Boil over high heat for 10 minutes. Turn to low heat and simmer for 3 hours.
2. Taste the soup before seasoning with salt.

soup

Lean pork soup with Tu Fu Ling, dried figs and oysters
_ 土茯苓無花果蠔豉湯 P.98

INGREDIENTS:

300 g lean pork # blanched in boiling water; drained
75 g Tu Fu Ling # soaked in water till soft
225 g dried oysters # soaked in warm water till soft
150 g dried mussels # soaked in warm water till soft
75 g laver # soaked in water briefly to remove the sand
10 dried figs
4 slices ginger
18 cups water
salt # to taste; added last

METHOD:

1. Put all ingredients (except salt) into a pot. Bring to the boil and cook over high heat for 10 minutes. Turn to medium-low heat and cook for 3 hours.
2. Taste the soup before seasoning with salt.

GIGI'S NOTES:

This soup alleviates haemorrhoid and eases constipation.
I prefer big dried figs from the U.S. for this recipe.
I used dried Tu Fu Ling in this soup. It expels Dampness and detoxifies. You can get it from Chinese herbal stores.

soup

Lean pork soup with monkey head mushrooms
_ 猴頭菇瘦肉湯 P.100

INGREDIENTS:

300 g lean pork # blanched in boiling water
75 g dried conch # see p.159 for preparation
75 g dried monkey head mushrooms # soaked in water till soft
75 g dried lily bulb # soaked in water briefly
38 g Dang Shen # rinsed
19 g Bai Zhu
4 candied dates
1 whole dried tangerine peel # soaked in water till soft; with pith scraped off
18 cups water
salt # to taste; added last

METHOD:

1. Put all ingredients into a pot (except salt). Boil over high heat for 10 minutes. Turn to medium-low heat and cook for 3 hours.
2. Taste the soup before seasoning with salt.

Gigi's notes
This soup helps prevent gastric reflux.
Monkey head mushrooms are sweet in taste and neutral in medicinal nature. It benefits the Spleen and Stomach while aiding digestion. Bai Zhu helps expel Dampness.

soup

Old cucumber soup with chicken feet and burdock
_ 老黃瓜養陰湯 P.102

INGREDIENTS:

1 large old cucumber # rinsed; de-seeded; cut into chunks with skin on
600 g burdock # peeled; cut into chunks
150 g dried conch # see p.159 for preparation
8 chicken feet # blanched in boiling water
38 g peanuts # rinsed; soaked in water briefly
38 g dried chestnuts # shelled; peeled
38 g lotus seeds # rinsed; soaked in water briefly
75 g sweet and bitter almonds
8 red dates # crushed; stoned
20 cups water
salt # to taste; added last

METHOD:

1. Put all ingredients (except salt) into a pot. Boil over high heat for 10 minutes. Turn to medium-low heat and cook for 3 hours.
2. Taste the soup before seasoning with salt.

GIGI'S NOTES:

This soup is best served in early summer as it helps expel Dampness and nourishing the Yin. The chestnuts also help secure Jing (essence of life) in the Kidney meridians.

soup

Lean pork soup with Tu Fu Ling and Mirabilis tuber

_ 土茯苓鑽地老鼠瘦肉湯 P.104

INGREDIENTS:

300 g lean pork # blanched in boiling water, drained
75 g dried conch # see p.159 for preparation
75 g Tu Fu Ling # soaked in water for 2 hours; rinsed
38 g dried Mirabilis tuber # rinsed
38 g raw Job's tears # rinsed
38 g puffed Job's tears # rinsed
75 g Huai Shan # rinsed
1/2 cup lotus seeds # rinsed
4 candied dates
1 whole dried tangerine peel # soaked in water till soft; with pith scraped off
18 cups water
salt # to taste, added last

METHOD:

1. Put all ingredients (except salt) into a pot. Boil over high heat for 10 minutes. Turn to medium-low heat and cook for 3 hours.
2. Taste the soup before seasoning with salt.

GIGI'S NOTE:

Raw Job's tears are Cold in nature. Puffed Job's tears are less Cold and help neutralize the Coldness of the raw ones. On the other hand, pearl barley is a different grain with a gooey texture after cooked. Pearl barley is mostly used in sweet soups.

Mirabilis tuber clears Heat, detoxifies, promotes diuresis and eases oedema. From Chinese medical point of view, body odour and smelly feet are caused by Dampness trapped in the body. Soups made with Mirabilis tuber help expel Dampness.

You can get fresh Mirabilis tuber from the wet market, or dried ones from Chinese herbal stores. I used dried one for this recipe. If you prefer using fresh Mirabilis tuber, you have to put in one whole tuber.

soup

Crucian carp soup with burdock_ 牛蒡鯛魚湯 P.106

INGREDIENTS:

1 crucian carp # dressed and rinsed; wiped dry; fried in oil till golden on both sides
1 burdock # peeled; cut into pieces
2 ears sweet corn # each cut into quarters lengthwise
3 carrots # peeled; cut into chunks
38 g lotus seeds # rinsed
38 g cashew nuts # rinsed
38 g Goji berries # rinsed
75 g dried lily bulb # rinsed
18 cups water
salt # to taste; added last

METHOD:

1. Put the fried fish into a muslin bag and tie tightly. Put all ingredients (except salt) into a pot. Boil over high heat for 10 minutes. Turn to medium-low heat and cook for 3 hours.
2. Taste the soup before seasoning with salt.

GIGI'S NOTES:

Lily bulb clears Fire, eases congestion in the Liver meridian and relieves bad mood. This soup regulates the Heart meridian, calms the nerves and eases stress.
I always put the fish into a muslin bag whenever I make fish soup. You can rest assured that you won't choke on any straying fish bone.
Frying the fish before making soup with it helps remove the fishy taste.

soup

Pork pancreas soup with pork small intestines, and Ji Gu Cao_ 豬橫脷粉腸雞骨草湯 P.108

INGREDIENTS:

300 g pork small intestines
2 pork pancreases # with fat trimmed off; blanched in boiling water
300 g Ji Gu Cao # rinsed
1 whole dried tangerine peel # soaked in water till soft; with pith scraped off
8 candied dates
18 cups water
salt # to taste; added last

METHOD:

1. Peel a clove of garlic. Stuff it into one end of the pork small intestine. Squeeze with fingers to run the garlic from one end of the intestine to the other. Repeat one more time to get rid of any dirt inside. Coat lightly in caltrop starch. Rinse well. Blanch in boiling water. Drain.
2. Put all ingredients (except salt) into a pot. Boil over high heat for 10 minutes. Turn to medium-low heat and cook for 3 hours.
3. Taste the soup before seasoning with salt.

GIGI'S NOTES:

This soup clears Heat; detoxifies; eases congestion in Liver meridians; alleviates blood stasis; gets rid of the bitterness in the mouth and eases hot flashes with restlessness.
Before rinsing Ji Gu Cao, soak it in water for 4 hours to remove the sand and dirt. It usually comes with much dirt and impurities.

soup

Lean pork soup with dried conch and Shi Hu _ 螺頭石斛湯 P.110

INGREDIENTS:

38 g Shi Hu # rinsed
150 g dried conch # see p.159 for preparation
300 g lean pork # blanched in boiling water, drained
75 g Huai Shan # rinsed
75 g Tu Fu Ling # rinsed
38 g Dang Shen # rinsed
38 g lotus seeds # with skin on, rinsed
38 g Goji berries # rinsed
1 whole dried tangerine peel # soaked in water till soft, with pith scraped off
20 cups water
salt # to taste; added last

METHOD:

1. Put all ingredients (except salt) into a pot. Boil over high heat for 10 minutes. Turn to medium-low heat and simmer for 3 hours. Serve.
2. Taste the soup before seasoning with salt.

comments...

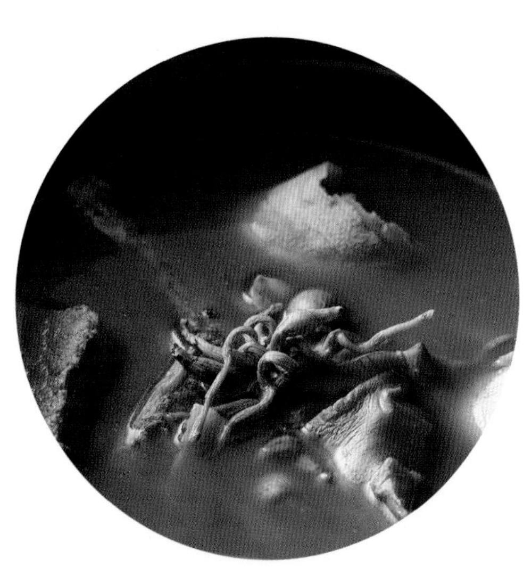

Stir fried pork belly with yam bean

soup

Pork bone soup with Chinese marrow, dried octopus and black-eyed beans_ 節瓜章魚眉豆湯 P.112

INGREDIENTS:

150 g dried octopus # soaked in water till soft
300 g pork bones # blanched in boiling water, drained
2 large Chinese marrows # peeled, each cut into chunks
1 cup black-eyed beans # soaked in water briefly
1 cup peanuts # with skin on, soaked in water briefly
2 slices ginger
18 cups water
salt # to taste; added last

METHOD:

1. Put all ingredients (except salt) into a pot. Boil over high heat for 10 minutes. Turn to medium-low heat and simmer for 3 hours.
2. Taste the soup before seasoning with salt.

GIGI'S NOTES:

This soup strengthens Qi (vital energy), promotes secretion of body fluids, refreshes and energizes the brain. It is useful for those with nervous prostration, general exhaustion and physical weakness.

Do not remove the pink skin of the peanuts. The soup is effective in regenerating blood cells only with both the skin and kernels of the peanuts.

No matter you use dried octopus for soup or other dishes, it's advisable to peel off the thin membrane on the surface after re-hydrating the dried octopus. Apart from getting rid of the acrid taste, the octopus flavours also seem to be infused more easily this way.

Chinese marrows turn mushy and break down easily after prolonged cooking. It's important to keep them in big chunks for them to stay in one piece.

Gigi Wong #This is a recipe from the famous pop diva Paula Tsui that demonstrate the beauty of simplicity. The four ingredients are easily available, but the sweet soup tastes extraordinary. The key is to stir in a knob of butter at last. You may use unsalted one, but salted butter tastes better.

comments...

sweet

Sweet soup in Paula's style_ 小鳳姐糖水 P.114

INGREDIENTS:

2 cups sweet corn kernels, 4 cups milk, 1/2 cup sugar, 1 small cube butter

METHOD:

1. Put sweet corn, milk and sugar into a pot. Cook over medium-low heat until sugar dissolves. Make sure you keep stirring it as it tends to boil over easily.
2. Stir in butter right before turning the heat off. Serve.

sweet

Glutinous rice balls with peanut butter filling in ginger syrup_ 薑片花生湯圓 P.116

INGREDIENTS:
1 cup glutinous rice flour
2/3 cup water
peanut butter # rolled into 12 balls; kept in freezer till set

GINGER SYRUP:
10 slices ginger
1 raw cane sugar slab
2 cups water

METHOD:
1. In a mixing bowl, put in glutinous rice flour. Add water little by little while kneading into dough. Cover with damp towel for 30 minutes.
2. Put all ginger syrup ingredients into a pot. Cook over medium heat until sugar dissolves and the syrup is infused with ginger flavour.
3. Roll the dough from step 1 into small balls. Press to flatten. Wrap a peanut butter ball in the dough. Seal the seams and roll it round. Repeat this step until all ingredients are used. Cook in the boiling ginger syrup until they float. Serve.

GIGI'S NOTES:
You may use any filling you like, such as red bean paste, sesame paste, diced cane sugar, or shredded dried coconut with sugar.
I freeze the peanut butter first so that it's easier to wrap the dough around it.

Horlicks cookies_ 好立克曲奇 P.118

INGREDIENTS:
3/4 cup self-raising flour
3/4 cup Horlicks # sieved
150 g unsalted butter # at room temperature; or chilled and heated in microwave for 20 seconds
1 egg # whisked
1 tbsp vanilla essence
1/2 cup nibbed almonds

METHOD:
1. Put all ingredients (except nibbed almonds) into a mixing bowl. Beat over low speed for 2 minutes to mix well.
2. Add nibbed almonds. Knead to mix evenly. Roll the dough into a log and wrap it in a sheet of baking paper. Refrigerate for 2 hours till stiff.
3. Slice the log of dough into thin slices. Arrange on a cookie sheet lined with baking paper. Make sure you leave some space between the cookies.
4. Preheat an oven to 170°C. Bake the cookies for 10 minutes. Flip them and bake for 8 more minutes. Remove from oven and let cool completely. Store in airtight jars for consumption later.

GIGI'S NOTES:
You may use 3-in-1 instant coffee or mocha coffee granules in place of Horlicks.
You may use sesames instead of nibbed almonds.

sweet

Sesame peanut brittle_ 芝麻花生糖 P.121

INGREDIENTS:

300 g white sesames # rinsed; drained
150 g peanuts # with pink skin on; rinsed; drained
8 tbsp white sugar
1 cup maltose

METHOD:

1. Fry white sesames and peanuts separately in a dry wok over medium-low heat until peanuts are lightly browned and crispy. Peel them.
2. Mix together sesames and peanuts.
3. In a non-stick pan, heat white sugar over low heat until it begins to melt and turns lightly browned. Turn off the heat. Add maltose and stir quickly until it thickens. Add half of the peanut and sesame mixture to the sugar mixture at one time. Mix well after each addition.
4. Pour the resulting mixture into a tray lined with baking paper. Press to flatten. Let cool. Cut or break into pieces. Serve.

GIGI'S NOTE:

If you prefer a neater outline instead of a rugged rustic look, cut the peanut brittle into your favourite shapes before it cools and sets.

Sweet potato cornbread_ 番薯玉米餅 P.124

INGREDIENTS A:
675 g sweet potatoes # rinsed; pierced with a fork all over
110 g butter # at room temperature

INGREDIENTS B:
4 eggs # whisked
1/2 cup milk

INGREDIENTS C:
1/2 cup self-raising flour
1.5 cups cornmeal
1/4 cup sugar
1 tsp salt
2 tsp ground cinnamon

METHOD:
1. Heat sweet potatoes over high power in a microwave oven for 5 minutes. Flip them over and heat them again over high power for 5 minutes. Peel and stir in butter while still hot.
2. Pour ingredients B into the sweet potato. Beat well with an electric mixer.
3. Mix all ingredients C. Pour half of the mixture into the sweet potato mixture from step 2. Stir well. Add the remaining half of ingredients C. Mix again. Pour into a baking tray.
4. Preheat an oven to 220°C. Bake the cornbread for 45 minutes. Serve.

GIGI'S NOTES:
This recipe calls for two fibre-rich ingredients, namely cornmeal and sweet potato. You don't need to mash the sweet potato too fine. It tastes even better with chunky bits at times.
Try to get sweet potatoes of similar thickness and size. It's easier to cook them through at the same time in the microwave oven this way.

sweet

Coconut rum panna cotta _ 椰香奶凍 P.126

INGREDIENTS:

100 g canned coconut milk
20 g rum
2 eggs # whisked
1 tsp gelatine powder
60 g sugar
500 g whipping cream # beaten with electric mixer till stiff

METHOD:

1. Pour coconut milk into a mixing bowl. Add rum, eggs, gelatine and sugar. Put in the whipped cream. Beat with an electric mixer till well incorporated.
2. Divide the mixture among small serving bowls or cups. Refrigerate for 1 hour. Serve.

GIGI'S NOTES:

This pudding is creamy and velvety in texture while bursting with coconut flavour. It is the perfect way to quench summer heat.

鳴謝 _SPECIAL THANKS ▽

謝謝您們 _ 讓這食譜的內容更豐富！

徐小鳳

徐小鳳？！還需要介紹嗎？

只要是香港人，都知道徐小鳳是殿堂級的歌星，是香港的 icon ！

只要是見到圓點，都會聯想到徐小鳳……舞台上她穿著的白地黑點長裙裏跳出十幾個小朋友，創出經典的一幕！

從此，只要見到圓點，大家都會會心微笑，脫口而出：小鳳姐！

我們識於微時，各自忙於自己的工作崗位。

縱使幾年才聚一次，我們都有說不完的話題！

欣賞小鳳姐的俠義精神，更欣賞她的體貼柔情！

有幸嘗過她自創的糖水，無以名之，直呼「小鳳姐糖水」，願與大家分享！

圖圖

圖圖，一個帶着濃厚古典美的現代潮人，藝術家，我的時裝店拍檔！

她的陶瓷創作、她的畫風、她的室內設計……都讓我知道，簡單就是美！

圖圖亦是廚藝高手，她要求很高，每一碟菜都彷彿是一件藝術品！一件雕塑！美得讓人不忍心吞入肚！

我把她的「洋葱鴨」簡單化，讓你收納在宴客菜單內！

吳政栓中醫師

吳政栓中醫師，是我的醫藥食療活字典！

每次寫食譜，一涉及藥材用料，即使常用的蓮子、百合、淮山……想要準繩地講出它的功效及配搭，定必急電吳醫師，有了他的指導、分析才可安心落筆！

也因此，我的湯水食譜寫得頭頭是道，非常權威！以致，被誤解我的醫學常識豐富，甚至要我開藥方，幫忙救人哩！

吳醫師，你幫了我，也「害」了我啊！

陳國強

常說在彌敦道叫一聲：「國強」！肯定有幾十人回頭，若然叫「陳國強」！回應何止上百！

陳國強，就是一個這麼普通的名字！

然而今天我為大家介紹的這位，卻是廚藝絕不普通的大廚陳國強！

在「吾淑吾食」的合作過程中，得到他諄諄指導，無私的分享！獲益良多！

他做的菜，精緻中特顯超凡！

更令我敬佩的，是他對太太的尊重與呵護！

粗中有細，不單是陳師傅的菜式，也是他給我的印象！

鄧達智

鄧達智對事物的敏銳觸角，往往令我目瞪口呆！

作為時裝設計師，一塊尼泊爾織錦，不需動剪，只要幾針，便成一件獨特的披風！

作為「吾淑吾食」的嘉賓，對食材的來源、運用都瞭如指掌！

涉足飲食界，將媽媽的味道，圍村的飲食文化發揚光大！

有鄧鄉紳的支持與鼓勵，何其幸運！

隨心。煮意

Cook it My Way

作者
黃淑儀

Author
Gigi Wong

編輯

Project Editor
Catherine Tam

攝影

Photographer
Imagine Union

美術統籌及設計

Art Direction & Design
Amelia Loh

排版

Typesetting
Sonia Ho

出版者
香港鰂魚涌英皇道1065號
東達中心1305室
電話
傳真
電郵
網址

Publisher
Forms Kitchen
Room 1305, Eastern Centre, 1065 King's Road,
Quarry Bay, Hong Kong.
Tel: 2564 7511
Fax: 2565 5539
Email: info@wanlibk.com
Web Site: http://www.wanlibk.com
http://www.facebook.com/wanlibk

發行者
香港聯合書刊物流有限公司
香港新界大埔汀麗路36號
中華商務印刷大廈3字樓
電話
傳真
電郵

Distributor
SUP Publishing Logistics (HK) Ltd.
3/F., C&C Building, 36 Ting Lai Road,
Tai Po, N.T., Hong Kong
Tel: 2150 2100
Fax: 2407 3062
Email: info@suplogistics.com.hk

承印者
百樂門印刷有限公司

Printer
Paramount Printing Company Limited

出版日期
二零一七年七月第一次印刷

Publishing Date
First print in July 2017

Published in Hong Kong by Forms Kitchen,
a division of Wan Li Book Company Limited.
ISBN 978-962-14-6382-1